AUSTIN TRACTORS

NICK BALDWIN

AMBERLEY

First published 2017

Amberley Publishing
The Hill, Stroud
Gloucestershire, GL5 4EP

www.amberley-books.com

British Library Cataloguing in Publication Data.
A catalogue record for this book is available from the British Library.

ISBN 978 1 4456 6828 4 (print)
ISBN 978 1 4456 6829 1 (ebook)

Typeset in 10pt on 13pt Sabon.
Origination by Amberley Publishing.
Printed in the UK.

Acknowledgements

In forty years of researching vehicle company histories, that of Austin and its tractors has puzzled and intrigued me the most and I'm grateful to the following friends and colleagues for their invaluable assistance:

Julie Baldwin, Roger Bateman, Roger Desborough, Stuart Gibbard, Brian Heath, Alain Bourgignon and his group Les Amis de l'Histoire based at the Mairie 60290 Cauffrey, J-M. Lecuru, J-P. Naninck, Nick Moffat, Yves Robineau, the Centre Historique Arts et Metiers, 60140 Liancourt, Neil Tuckett, Chris Barker and Mike Worthington-Williams. Much of the background to the British Austin Motor Co. was helped by the previous research of Nick Georgano and the late Z. E. Lambert, as well as R. J. Wyatt and Michael Sedgwick.

Contents

Chapter 1

Austin and America

After almost two years of spectacular growth, the Austin Motor Co. entered the First World War with 2,638 employees and a production of well over a thousand cars and lorries per year, plus petrol-powered electrical generators.

Austin's factory at Longbridge converted rapidly to war work after Herbert Austin's only son, Vernon James, was killed by a sniper on the Western Front on 26 January 1915. He was a twenty-one-year-old Second Lieutenant in the Royal Field Artillery and would have been destined to take over the factory.

By the war's end there were 22,000 men and women working on Longbridge's 58 acres and output had included 350 armoured cars, 10,000 switchboards, 2,000 aeroplanes, 2,500 aero engines, 150 ambulances plus 750 car chassis for assorted roles, 3,700 trailers for various purposes including gun limbers, 650 howitzers, 4,762 generator sets and 8 million shells. There were also 2,000 lorries equipped for such diverse roles as trench pumps, field hospitals, kitchens, searchlights and troop carriers. Government munitions factories erected on nearby farmland later became Austin's North and West Works. Herbert Austin employed 3,000 Belgian refugees and for this was awarded Commandeur de l' Ordre Leopold II by the Belgian king, as well as the Croix de Guerre from the French for ambulances and other support.

German success at destroying American war material destined for the Allies had the knock-on effect of bringing Britain close to starvation as food convoys were sunk in the Atlantic. The immediate response in Britain was to bring more land into agricultural production but, with so many men and horses called up for battle, there was a desperate need for mechanical assistance.

Farm tractors had been known in America and parts of Europe ever since automobiles had started to make an impact around 1900. Britain's Ivel and Saunderson were among the pioneers, but sales were poor due to high prices, doubts about new-fangled machinery, and a surplus of labour and horses. Both firms plus many other hopefuls, including the makers of steam traction and ploughing engines, were producing tractors in the teens, but output was far too small to make a significant impact and, in consequence, Britain turned to America. Here hundreds of manufacturers had entered the far larger market, including significant agricultural machinery makers like International Harvester and JI Case Plow Works (there was also a JI Case Threshing Machine Co. making Case tractors).

The JI Case Plow Works President, H. M. Wallis, was involved with the archetypal Wallis Cub tractor of 1913. Whereas all other types were effectively self-propelled stationary farm engines mounted on girders and wheels, the Cub had a monocoque torsionally rigid curved frame, to which all mechanical components were attached for strength and perfect alignment.

There was still a place for the primitive type, as the Bull Tractor Co. of Minneapolis proved when it mass produced, at a low price, a machine that sold almost 4,000 examples between April and December 1914. A rather similar Waterloo Boy appeared that year and around 8,000 examples had been sold by 1919, the year after plough-maker John Deere had acquired the business. These and numerous other types of tractor dodged the enemy blockade to save Britain, and Herbert Austin took a leaf from their book by importing his own selection.

Austin's new Agrimotor department manager, Walter L. Bodman, represented the Austin Motor Co. and the British government when he went to America in autumn 1916 to study the latest trends and pick some suitable makes for sale in Britain. At the basic end of the range, a machine with single driven wheel, side balance wheel and steered front wheel was chosen from the Peoria Tractor Co. of Peoria, Illinois (a town later to be famous for the Caterpillar tractor). At $685 it cost $100 more than the broadly similar 1916 Big Bull and became known as the Austin Culti-Tractor Model 1, which, like the subsequent American models selected for Austin, carried a six month guarantee and, in many cases, free delivery and training.

On 5 October 1916 a contract was signed with barbed wire magnate Joseph M. Denning of Cedar Rapids, Iowa granting Austin British, Irish and Continental sales rights for his recently introduced Denning tractors. At a price of $850 each, plus $40 for crating and shipment with a twenty-five per cent discount, 300 were in the initial order. The idea was to trial them first in Warwickshire and Worcestershire and then order a further 350, followed by 400 more.

Also in October 1916, Bodman agreed to take 300 of a rather similar-looking fully enclosed (but still primitively built) girder machine from the Interstate Engine and Tractor Co. of Waterloo, Iowa, and this became the Austin Culti-Tractor Model 2. He negotiated a twenty-two per cent discount but, no doubt in competition with Denning, he managed to increase this to twenty-five per cent.

For some reason, the Denning did not enter the Culti-Tractor numbering system and nor did the next batch, which was seventy-five Joliet oil tractors made in the town of the same name in Illinois. This was a curious machine controlled from the towed implement and incorporating a single driven track but steered front wheels. The order of 7 October 1916 covered a sample tractor at $760 plus $75 for crating and $150 for spares (which perhaps implied various known defects like track and sprocket wear). If all went well, an order for 300 would follow, and an optimistic clause spoke of taking twenty-five per cent more machines in each subsequent order and spending $6,000 on advertising – half of which would be reimbursed by Joliet. Some of these Joliets were intended for former Autocar editor Henry Sturmey, who had entered the motor business with his Lotis vans before the war.

On 26 January 1917, Austin added a fifth line as its Culti-Tractor Model 3 with HM Strait's Killen-Strait, made in Appleton, Wisconsin. This also had a single driven track at the rear, but instead of a steered front wheel had a free-running track that could be pointed in the direction of travel by a conventional steering wheel. The contract spoke of twenty-five machines at cost price plus $10, plus spares, plus ten per-cent, making a grand total of $1,200. However, this price came down by a further twenty per-cent for 250 delivered within one year and mention was made of 300 to 350 thereafter.

It would be interesting to know how many Culti-Tractors were sold and how successful they turned out to be. By March 1917, government figures quoted 600 US tractors of all types ready for work in Britain. One man who would have known was Harry Ferguson, who helped the government keep several makes working throughout Ireland. He travelled continuously by car and on numerous occasions slept in it by night. Ferguson would be an Austin agent in the 1920s and an important ally. In the 1930s he would, of course, develop his own highly significant tractor.

One of several American tractor makers not encountered by Walter L. Bodman was FC Austin of Chicago, Illinois. Frederick Carlton Austin had been making construction and earthmoving machinery since 1888. In 1902 he had merged with the Western Wheeled Scraper Company and, as Austin-Western, became the world's largest (pre-Caterpillar) maker of earthmoving and construction equipment. The 20-acre FC Austin factory in Chicago also operated under its own name into the 1930s and from about 1918 had made wheeled and crawler farm tractors. Production survived only a few years, though the crawler bulldozer and excavators proved to be more successful and some wheeled Austin-Western tractors were supplied to Britain for aircraft runway construction during the Second World War. At the end of the First World War, the British Austin company had tried to take out an injunction to stop the American firm from using the Austin name, but had no success.

However, this was the least of Sir Herbert's problems (he was knighted in 1917 for war services), as Henry Ford & Son registered a tractor design on 27 July 1917 at Dearborn, Michigan, that was about to take the entire farming community and tractor industry by storm. This was due to its modern design on the lines of the integral construction Wallis (though with cast rather than rolled frame). It undercut virtually all competition on price due to mass production.

To make matters worse for Sir Herbert's potential sale of his American machines, Henry Ford offered its plans free of charge to the British Ministry of Munitions, who called it the MOM tractor. The idea was for British manufacturers to copy the design and thus outwit the German Navy. The MOM tractor project was put under the control of Percival Perry of Ford's Manchester assembly plant and S. F. Edge, a farmer, motor industry tycoon and car racer, who had been an early convert to the Ivel tractor.

It was soon obvious that British industry was stretched to the limit by its war work and that the naval blockade was beginning to fail, so 5,000 tractors were ordered direct from Dearborn, soon increased to 6,000 with many arriving in 1917 and the rest by April 1918. They were sold at cost price of $700 plus a modest profit of $50 per tractor. Ford's total American tractor output in 1918 totalled 34,167, which put it in first place ahead of International Harvester with most of the hundreds of other rivals en route to oblivion.

All of Sir Herbert's imports were to succumb but with the MOM he had seen the writing on the wall and moved the rights to most of them to the Vulcan Car Agency in London on 11 June 1917. He had realised that the Ford design would succeed and that there was no sound business reason to make it himself, though there was nothing to stop him making an improved version that did not infringe any Ford copyright. From 1917 he began to apply for numerous patents relating to such details as front axle attachments, connecting rods for vee engines (not actually employed in his tractors) and reinforced driving wheels. In May 1919, Austin placed advertisements in the farming press to clear the last of his stock of Interstate, Killen-Strait and Bates Steel Mule (the new name of the Joliet in honour of the firm's vice president, Harry H. Bates). Presumably the Vulcan Car Agency had only taken the ones it could sell as the Austin Motor Co. still had a large stock that, it patriotically claimed, had been bought 'in the interests of agriculture and the national emergency ... now offered for sale at greatly reduced prices in order to clear the way for post-war products'. The Culti-Tractor name had been shelved by then, which was just as well as the United Tractors Corp of New York had picked it, but without the hyphen, for its new Mohawk Cultitractor. This had only a brief existence in view of Ford's saturation of the embryonic tractor market.

The mention of Ford once more brings us to the remarkable parallel careers of Henry, born 1863 in Michigan, and Herbert, born three years later in Buckinghamshire, both into farming backgrounds. Both achieved success through mass production and both never forgot their roots and made efforts to ease the physical burden of farmers through mechanisation.

The Peoria, or Austin Culti-Tractor Model 1, being shown to motor dealers, including George Heath in the Midlands.

THE VULCAN CAR AGENCY, LTD.

THE "CULTI-TRACTOR"

BIG POWERED—SIMPLY CONSTRUCTED—EASILY OPERATED.

FOR MODERATE-SIZE FARM AND ORDINARY LAND CONDITIONS.

This model represents the best value ever offered in the Tractor Market. It will plough 3 furrows in any ordinary pasture or arable land, or 2 furrows on hills, or in heavy wet land. Will operate all implements—Disc and Spike Harrows, Seeders, Mowers, Reapers, Binders, Threshers, outside Belt work of all descriptions—and do it all efficiently.

SHORT SPECIFICATION

MOTOR	4 Cylinders, 3¾in. bore × 5in. stroke, governed to a speed of 1,000 r.p.m.
TRANSMISSION	Spur Gearing—one speed forward and reverse—all high speed gears completely enclosed.
DRIVING WHEEL	5ft. 0in. high, 18in. wide, built up with turnbuckle spokes, fitted with angle strakes for land work.
STEERING	Bevel and quadrant to single wheel, carried in heavy fork—turning circle 20ft. 0in. —automatic steering device enables operator to leave Tractor when running furrows and follow and attend to plough or other implement.
FUEL	Start on Petrol, Paraffin used within 3 or 4 minutes after engine warms thoroughly —consumption 1 to 1½ gallons per acre.
SPEED	2¼ miles per hour, ploughing capacity 5 to 8 acres per day.
WEIGHT	Approximately 35 cwts.

PRICE [redacted] £285

Sales leaflet for the Culti-Tractor Model 1 after it had been adapted by the Vulcan Car Agency. Its Beaver engine's bore and stroke matched the subsequent Austin R.

Left: The Culti-Tractor logo of unknown colour adopted early in 1917 and faintly discernible on the rear wing of the Model 1.

Below: Interstate, known as Austin Culti-Tractor Model 2, had two gears forwards and backwards and could plough up to eight acres per day.

Above and below: Two views of the Interstate, which resembled the Denning but, unlike it, featured sun and planet exposed gearing on its five-foot-diameter rear wheels.

The 10–18 hp Denning Model E had a very upright driving position and the unusual feature of a coil-sprung front axle.

From being a barbed wire millionaire, Joseph M. Denning lost it all with his tractors and subsequent concrete fence posts. The tractor business was bought by General Ordnance of New York City in 1919 and, after a change of name to National then GO, it ended in 1922.

Culti-Tractor Model 3 was this £415 Killen-Strait that with £100 road pads could haul 5 tons on the road at its single forward gear maximum of 2.5 mph.

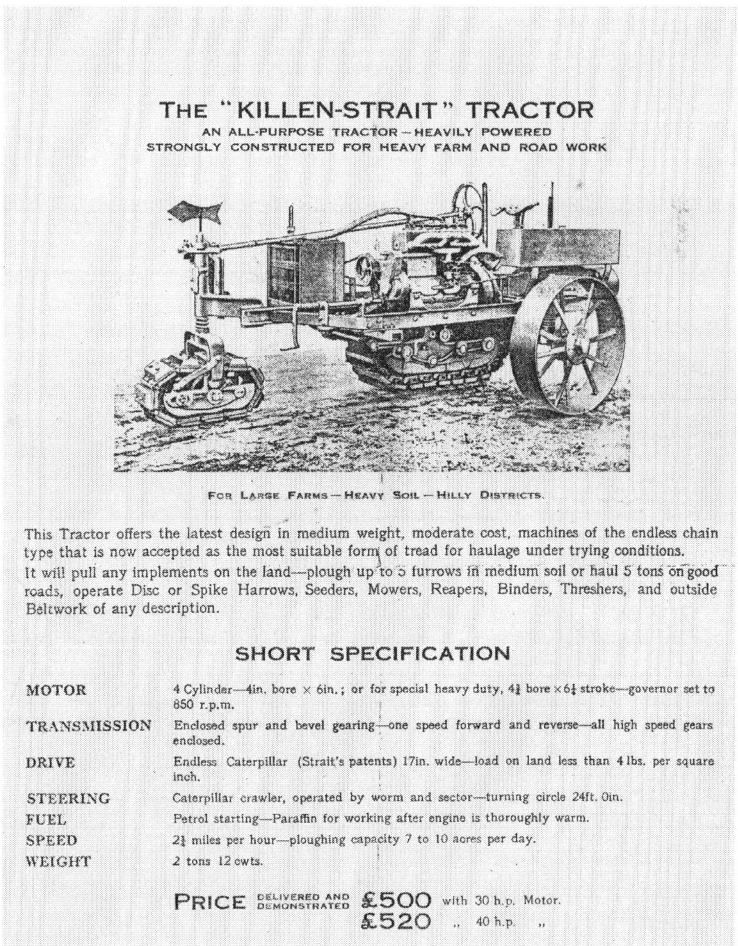

THE "KILLEN-STRAIT" TRACTOR

AN ALL-PURPOSE TRACTOR – HEAVILY POWERED
STRONGLY CONSTRUCTED FOR HEAVY FARM AND ROAD WORK

FOR LARGE FARMS – HEAVY SOIL – HILLY DISTRICTS.

This Tractor offers the latest design in medium weight, moderate cost, machines of the endless chain type that is now accepted as the most suitable form of tread for haulage under trying conditions.

It will pull any implements on the land—plough up to 3 furrows in medium soil or haul 5 tons on good roads, operate Disc or Spike Harrows, Seeders, Mowers, Reapers, Binders, Threshers, and outside Beltwork of any description.

SHORT SPECIFICATION

MOTOR	4 Cylinder—4in. bore × 6in.; or for special heavy duty, 4¾ bore × 6¼ stroke—governor set to 850 r.p.m.
TRANSMISSION	Enclosed spur and bevel gearing—one speed forward and reverse—all high speed gears enclosed.
DRIVE	Endless Caterpillar (Strait's patents) 17in. wide—load on land less than 4 lbs. per square inch.
STEERING	Caterpillar crawler, operated by worm and sector—turning circle 24ft. 0in.
FUEL	Petrol starting—Paraffin for working after engine is thoroughly warm.
SPEED	2½ miles per hour—ploughing capacity 7 to 10 acres per day.
WEIGHT	2 tons 12 cwts.

PRICE DELIVERED AND DEMONSTRATED £500 with 30 h.p. Motor.
£520 ,, 40 h.p. ,,

Sales leaflet for the Killen-Strait showing it from the balance wheel side.

Bates Steel Mule (formerly Joliet) at the Highland & Agricultural Society of Scotland's October 1917 trials.

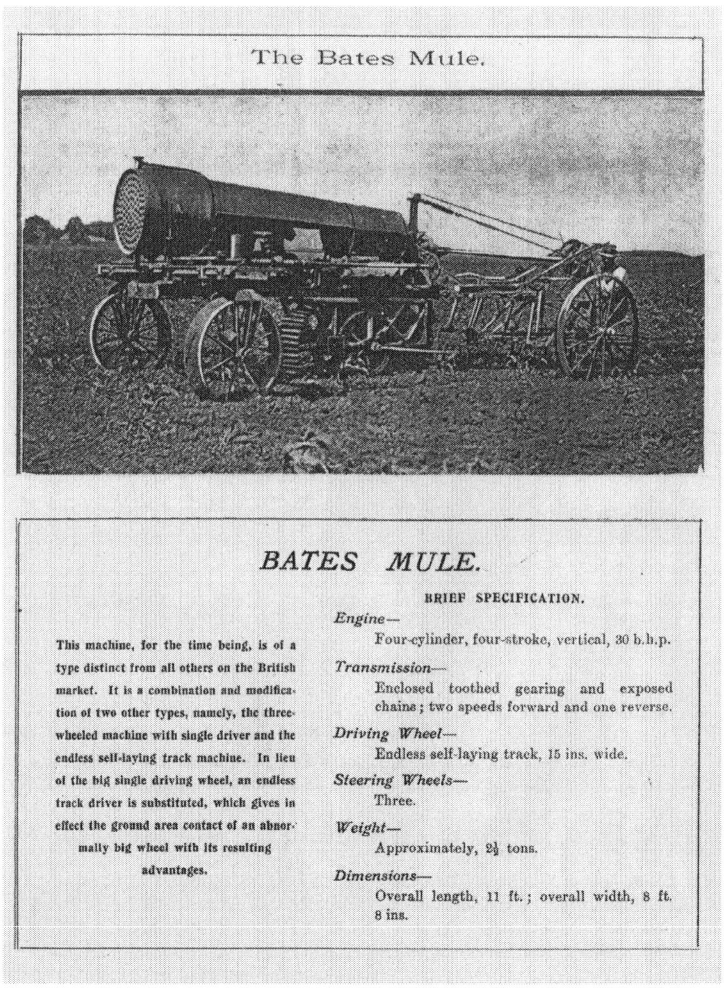

The Bates Mule.

BATES MULE.

This machine, for the time being, is of a type distinct from all others on the British market. It is a combination and modification of two other types, namely, the three-wheeled machine with single driver and the endless self-laying track machine. In lieu of the big single driving wheel, an endless track driver is substituted, which gives in effect the ground area contact of an abnormally big wheel with its resulting advantages.

BRIEF SPECIFICATION.

Engine—
Four-cylinder, four-stroke, vertical, 30 b.h.p.

Transmission—
Enclosed toothed gearing and exposed chains; two speeds forward and one reverse.

Driving Wheel—
Endless self-laying track, 15 ins. wide.

Steering Wheels—
Three.

Weight—
Approximately, 2½ tons.

Dimensions—
Overall length, 11 ft.; overall width, 8 ft. 8 ins.

Sales leaflet for the Bates showing it being controlled remotely from the implement by a driver with very restricted forward vision.

Austin
TRADE MARK
GREAT DEMONSTRATION
OF
FARM TRACTORS
at Northfield, near Birmingham,
on Thursday, 22nd, & Friday, 23rd November.
THE DEMONSTRATION WILL INCLUDE
HAULING of PLOUGHS and show other
AGRICULTURAL MACHINERY
AT WORK ON THE LAND. . . .
ALSO
CHAFF-CUTTING, CORN-CRUSHING & THRESHING
MACHINES IN ACTION.
YOU ARE INVITED TO ATTEND.
Admission Tickets free to all interested, supplied on application to
TRACTOR DEPT.,
Austin Motor Co., Ld.,
NORTHFIELD, BIRMINGHAM.
Telegrams: Tractor Dept., "Speedily, Northfield."
Telephone: King's Norton 230 (Tractor Dept.).
IMPORTANT NOTE.—Apply early in order that adequate transport and catering facilities may be provided.

November 1917 publicity for Austin's own tractor trials that took place near its factory, where plans were already in place for a British tractor.

Plans for the MOM tractor of Ford origin were available for inspection in London. Those wishing to copy them had until 2 June 1917 to submit their intentions.

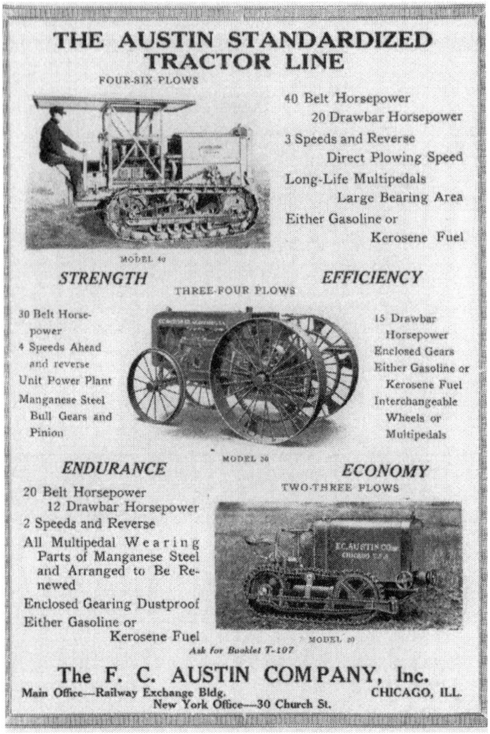

Sir Herbert Austin's rival in Chicago advertises a short-lived range of farm tractors. FC Austin acquired the Wilson 4x4 tractor business in 1922, though this and the crawlers were primarily for earthmoving.

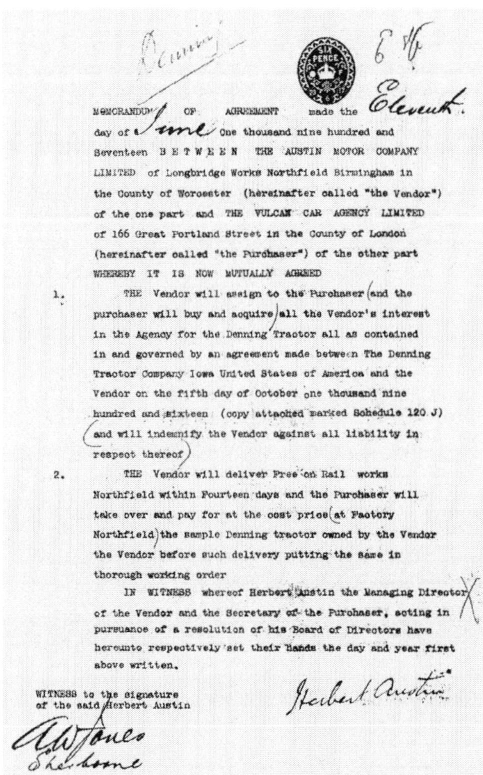

Denning

SIX PENCE

6 W

MEMORANDUM OF AGREEMENT made the Eleventh day of June One thousand nine hundred and Seventeen BETWEEN THE AUSTIN MOTOR COMPANY LIMITED of Longbridge Works Northfield Birmingham in the County of Worcester (hereinafter called "the Vendor") of the one part and THE VULCAN CAR AGENCY LIMITED of 166 Great Portland Street in the County of London (hereinafter called "the Purchaser") of the other part WHEREBY IT IS NOW MUTUALLY AGREED

1. THE Vendor will assign to the Purchaser (and the purchaser will buy and acquire) all the Vendor's interest in the Agency for the Denning Tractor all as contained in and governed by an agreement made between The Denning Tractor Company Iowa United States of America and the Vendor on the fifth day of October one thousand nine hundred and sixteen (copy attached marked Schedule 120 J) (and will indemnify the Vendor against all liability in respect thereof)

2. THE Vendor will deliver Free on Rail works Northfield within Fourteen days and the Purchaser will take over and pay for at the cost price (at Factory Northfield) the sample Denning tractor owned by the Vendor the Vendor before such delivery putting the same in thorough working order

IN WITNESS whereof Herbert Austin the Managing Director of the Vendor and the Secretary of the Purchaser, acting in pursuance of a resolution of his Board of Directors have hereunto respectively set their hands the day and year first above written.

WITNESS to the signature of the said Herbert Austin

Herbert Austin

A W Jones
Sherborne

Contract signed by Herbert Austin handing over his interest in Denning tractors to the Vulcan Car Agency Ltd on the same date he assigned Interstate and Joliet (Bates) to Vulcan.

Chapter 2

Austin's Agricultural Roots

Herbert Austin was born at Grange Farm, Deep Mill Lane, Little Missenden, Buckinghamshire, on 8 November 1866. The family farm was impoverished and, when his father was offered the chance to be a farm bailiff, the family moved to the Wentworth Estate in Yorkshire. His brother was already there as Estate Architect to Earl Fitzwilliam, a coal magnate who, in 1906 (the year that the Austin Motor Company started production), bought the Brotherhood Crocker car firm from London and re-established it locally as Sheffield-Simplex. At the age of seventeen in 1884 Herbert went to Australia with an uncle who was works manager to engineers Mephan Ferguson in Melbourne. Herbert spent two years learning foundry and machining techniques before joining Cowans, who were agents for printing machines and Crossley gas engines from Britain.

He next worked on wheels, boilers and mining equipment at Longlands Foundry for the Ballarat gold fields and married a local girl of Scottish origins at the end of 1887. He went straight from his honeymoon to a workshop trying to perfect the mechanical sheep shears invented by Frederick York Wolseley, a Briton who had been running sheep stations since the 1860s. Wolseley soon invited Austin to join him as his engineer and in 1893 they returned to Britain to start precision manufacture in Birmingham. Herbert was a keen cyclist and, as well as sheep shears, made machine tools and bicycle parts and had a useful sideline selling Raglan and Rover bicycles to Australian friends.

Wolseley had a stand at the 1895 Paris International Exhibition, which is possibly where Herbert saw the Bollée three-wheel automobile made in Le Mans. Inspired, he produced a similar vehicle, followed by others, but received little encouragement from the Wolseley Sheep Shearing Machine Co. and from 1899 considered starting his own business.

Herbert Austin made components for machine-gun maker Hiram Maxim's steam aeroplane. Vickers, Son & Maxim Ltd took over Wolseley's 'autocar' interests early in 1901. Herbert agreed to run their Wolseley Tool & Machinery Co. for five years in exchange for £500 a year, a share allocation, and five per cent of profits. Its new factory was at Adderley Park in Birmingham on a 3.5-acre site formerly owned by Starley, a well-known name in the bicycle industry – one of whose members had started Rover. However, Herbert retained his links with the sheep shear business and acted as its chairman 1911–33. Wolseley autocar production rose from fifty in the first year to 190 in 1902, 270 in 1903 and 341 in 1904 – the year Herbert built his first tractor, a giant 24 hp twin horizontal cylinder machine, probably intended more for military than agricultural purposes. However, he felt neglected by Vickers, especially after it sold vertical-engined Peugeot-inspired cars in competition with his horizontal-engined types. Much of the growth in production, which leapt to 850 in 1905, was accounted for by these Siddeleys, named after their promoter John Davenport Siddeley. Before the five-year contract came up for renewal, Herbert Austin left (some say he was forced out) and started the Austin Motor Co. late in 1905 in White & Pike's derelict print works at Longbridge near Birmingham.

He took some Wolseley men with him, including his own brother Harry, and gained £20,000 backing from his old steel supplier, Frank Kayser in Sheffield. Much of the rest of the capital came from Harvey du Cros, who had acquired the Dunlop tyre patents in the 1890s and used the vast wealth these created to invest in the infant motor industry (a route also taken by Siddeley from the proceeds of his Clipper tyres). Harvey du Cros had a stand at the London Motor Show late in 1905 displaying Mercedes, but he found a corner where two ex-Wolseley draughtsmen showed drawings of the new Austins and even managed to take a few orders. In 1896 du Cros had bought out the French bicycle firms Clément and Gladiator and he went on to use Austin and Swift, another of his investments, to produce examples of their small cars. France was the largest exporter of cars to Britain and du Cros played a major part in financing and controlling its industry. His sons also built up the biggest fleet of taxi cabs in London and became significant makers of commercial vehicles.

Meanwhile, Austin's British production climbed from 120 in 1906, when 270 men were employed, to 200 cars in 1910 and 1,100 in 1912 (when the workforce stood at 1,800). Herbert Austin exhibited at Turin Motor Show and opened a Paris branch that came to offer cars, marine engines, house lighting plants and, from their introduction in July 1913, lorries. He entered racing car teams and power boats in Continental races, and became a frequent visitor to France, using its straight open roads for testing the latest models. He was plainly so impressed that after the war he bought the Sizaire-Berwick firm, which had made cars in France and Britain. It is chiefly remembered for employing *Dixon of Dock Green* actor Jack Warner, who taught Maurice Sizaire to speak English. Herbert Austin's daughter recalled that her father was a good linguist, though did not say if his French came with an Australian accent after his eleven years there.

Not far from Beauvais, en route from Dover to Paris, is the industrial town of Liancourt in the Oise region. It was here that Austin would establish his European factory in 1919, probably influenced by the skilled labour already working for agricultural engineers Albaret, Bajac and Tosello. The Albaret factory, founded in the adjoining town of Rantigny in 1847, was where Antoine Bajac was employed as a metal worker from 1867. He joined the plough-making business of Delahaye-Tailleur in 1871 and married the owner's daughter. As Machines Agricoles A Bajac, the factory prospered and grew to 1,300 metres in length, employing 230 men in 1905 and over 500 in the 1920s. At 50,000 square metres it was said to be the largest plough factory in France and, perhaps, the world – John Deere and Ransomes might have disagreed.

August Albaret had been among the early students of a revolutionary education system created at Liancourt. The Duke of Liancourt had founded the forerunner to France's

Arts et Metiers system (whose museum in Paris houses Joseph Cugnot's incredible 1769 military steam tractor). The first college was at the Ferme de la Montagne in 1780 and fifteen years later the Duke de la Rochfoucauld set up Liancourt's National School at his chateau. Nowadays, the Ferme de la Montagne houses an Arts et Metiers museum that includes a Bajac plough and Albaret steam roller.

The Duke de la Rochfoucauld had a model estate with cloth and saw mills, hat and shoe factories, and a brickworks. He had fled to Britain in the Revolution and became a friend of agronomist Arthur Young, who advised him of the latest agricultural trends for when he was able to return to France and develop his estates. Just as Industrial Revolution birthplace Ironbridge on the River Severn was an easy hour or two drive from Longbridge, Liancourt could claim to be the birthplace of the French equivalent, and may have been known to Herbert Austin before he bought the former leather, shoe and latterly Lorimer accumulator factory. This had been built in 1897 and was sometimes called Manufactures de Liancourt, situated next to the Bajac works with which it shared a common boundary.

While Albaret made steam traction engines from 1870, rollers from 1880 and direct steam ploughing engines from 1903, Bajac's first motor hoe of 1907 was petrol-powered, as was the winch plough unit made a year later. Tosello, founded in 1900, made stationary and portable engines for agricultural and industrial use, thus confirming that Liancourt had become a fertile seedbed for agricultural machinery development. The problem was that, although France was a massive farming nation, the size of average individual farms was tiny, so very few could afford the equipment.

The Maison Th. Pilter, founded in 1864, was a major distributor of agricultural implements, often of American origin, but in 1911 its motor tractors came from Ivel and its steam tractors and threshing machines from Garrett – both British manufacturers. Among its myriad other lines were hand-pumped vacuum cleaners from Austin's friend Hiram Maxim. In 1919 Pilter would become an Austin tractor distributor with sales branches all over France and its North African colonies.

Hiram Maxim returned to his native America to create the Maxim Munitions Corp in New York City. Turning swords into ploughshares towards the end of the First World War, this firm introduced a Maxim tractor, which was imported to Britain by a Birmingham motor dealer called Henry Garner, whose name appeared on lorries in the 1920/30s and on light tractors into the 1950s.

It has been widely noted that there was a real bond of affection between Herbert Austin and his staff. His daughter put it down to the very long hours he worked in the factory and his modest and affable manner (though he wouldn't tolerate laziness) – everyone knew him as Pa Austin. She reveals that he never smoked nor drank alcohol and worked long hours at home too, notably in the library of his newly acquired Lickey Grange. There he attended to paperwork and designs while listening to classical and opera music on gramophone records. He felt there to be a link between good music and good engineering – rhythms of machines, he felt, were directly connected to the mechanics of music.

Herbert Austin in bowler hat at the wheel of one of his early Austin cars.

Herbert Austin's first tractor was this horizontal twin cylinder 24 hp petrol-engined machine of 1904, built by Wolseley.

Antoine Bajac's plough works was the biggest in France and was situated next to Manufactures de Liancourt, which would become home to Austin tractors.

LES MANUFACTURES DE LIANCOURT
(Oise)
FABRIQUE DE CHAUSSURES
SOCIÉTÉ ANONYME
Statuts déposés chez Mr TANSARD, Notaire à PARIS
SIÈGE SOCIAL & USINE A LIANCOURT (OISE)
Action de 100 Francs au Porteur
ENTIÈREMENT LIBÉRÉE
Liancourt le 13 Août 1898

1898 share certificate of Manufactures de Liancourt showing its new factory. Bajac's works were left of the chimney.

Advertisement for a reversible Bajac plough, which allowed the adjoining furrow to be worked from each end of the field.

Bajac's motor hoe in 1910 had a twin-cylinder 8 hp engine and could be equipped with rake or seeder.

From 1909, Bajac offered hauled winches for ploughing and also winch tractors. It experimented with tractor-compressors with additional handheld tools worked by air.

Monument to the Duke of Rochfoucauld 1747–1827 at la Ferme de la Montagne in Liancourt, where industrial and artistic education went hand-in-hand.

1898 Albaret steam roller. A more modern version is preserved at the Arts et Metiers museum at la Ferme de la Montagne, Liancourt. Albaret had made steam traction engines since 1870.

ALBARET
Rantigny (Oise)
Succursale :
--- PARIS ---
7bis, rue du Louvre
Créations 1929
Batteuse V - 25 Quint.
Presse C - 12 Tonnes
Locomobile à huile lourde
10 CV.
DEMANDEZ LE CATALOGUE DE NOS
BATTEUSES - LOCOMOBILES - PRESSES à FOURRAGES

Albaret had a long history of making threshing machines. This advertisement dates from 1929 and speaks of the firm's 10 hp oil-engined portable power plant.

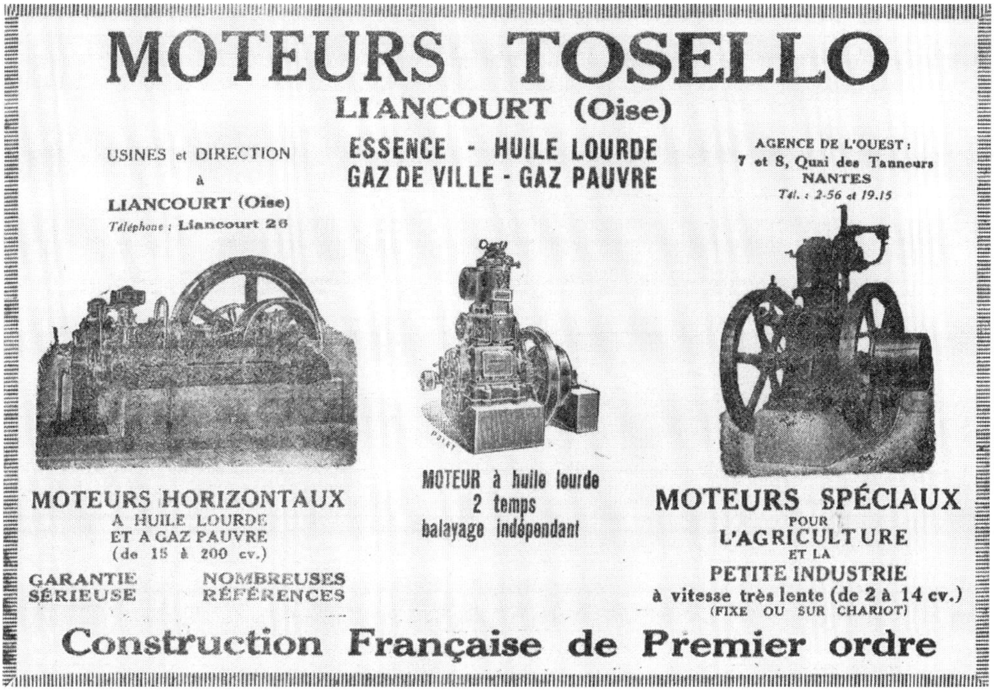

Liancourt was also home to this major maker of internal combustion engines for agriculture and industry.

Th. Pilter had branches all over France and her colonies, selling a wide range of agricultural equipment. In 1911 its steam range came from Richard Garrett of Leiston in Suffolk.

TRACTEUR AGRICOLE "IVEL"

PILTER

PARIS — 24, Rue Alibert, 24 — PARIS

La Maison PILTER est heureuse de pouvoir présenter à sa clientèle le premier tracteur agricole vraîment pratique, autant par sa facilité de conduite et d'entretien, que par l'économie réalisée par son emploi. Les services que peut rendre notre tracteur agricole sont innombrables ; pour en citer quelques-uns, on peut l'employer très avantageusement pour le labourage, la moisson, le battage, la traction sur route, etc., etc. Notre tracteur a déjà subi les épreuves les plus concluantes au point de vue de la facilité de manœuvre, même dans les terrains les plus difficiles.

La transmission du mouvement est faite au moyen de cônes d'embrayage et de chaînes ; la marche avant et la marche arrière sont commandées par un seul levier, le

Th. Pilter's only internal-combustion-engined tractor in 1911 was Britain's pioneering Ivel from Biggleswade, Bedfordshire. It was rated at 14 hp, cost 9,500 francs and weighed 1.4 tons.

The 1897 factory next to Bajac changed hands many times before it became Austin. Here it is called Les Autocommutateurs, making Lorimer batteries and electrical equipment.

A print works at Northfield, Longbridge, Birmingham, had been derelict for four years before Austin car production started there in 1906.

The Austin script drafted by Herbert Austin before the teens with the familiar winged wheel logo, both used on his tractors.

Monsieur Austin à bord "Maple Leaf IV." MOTEUR AUSTIN,
gagnant du BRITISH INTERNATIONAL TROPHY en 1912 et 1913.
Vitesse ; 49'02 noeuds (plus de 90 kms. a 1'heure).

Hands-on mechanic Herbert Austin in cap works on an Austin-engined powerboat. The photo is marked to rear Automobiles Anglaises Austin SA, 134 Avenue de Malakoff, Paris. Telegram Speedily, Paris.

Chapter 3

All Eggs in One Basket

The end of the First World War saw the cancellation of government orders, which left the vastly expanded Austin factory and its 22,000 employees with little to do. Sir Herbert Austin, a Unionist MP for his local constituency since 1918, had decided to follow Ford into a one-model policy plus aeroplanes. There would be a 20 hp car, a 20 hp lorry and the 20 hp tractor that he had refined from his study of the MOM plans and his experience of other American tractors. Using so many common parts across the range would give all the benefits of mass production and low unit costs.

However, it was soon plain that there was a downside to this plan: as post-war euphoria evaporated, the dream bubble of civilian flying burst and dealers discovered that their customers could not afford to buy, let alone run, large powerful cars. The previous Austin lorries had been unusual in featuring radiators on their scuttles and protruding coal scuttle bonnets, but were redesigned with front radiators identical to the tractors for 1920. Unfortunately, by then there were thousands of ex-War Department lorries for sale at low prices, and all makers of new lorries went through a long depression.

This left Austin's new tractor, which entered a market that scarcely existed. True, there had been wartime imports, but these had seldom been bought by individual farmers. There was no longer a dire shortage of draught horses and labourers, and lots of other British manufacturers had come up with new tractors, including steam engineers Ruston & Hornsby, who obtained a licence to build the revolutionary Wallis (see Chapter 1) at Lincoln. Most worryingly, there was the ultra-cheap Fordson from Dearborn and, from 1919, also Cork from where Henry's father had set sail for America in 1850. When US production ended in 1928, 740,000 had been sold. Cork operated to 1923 and then again from 1929 to 1932, making a total of 38,600 before production moved to England. Cork output had been hampered by early foundry difficulties and tariff restrictions in Britain following the formation of the Irish Republic.

The Fordson and Austin shared a very similar specification in 1919, though the latter had ignition by magneto rather than potentially unreliable trembler coil, and it had bevel and spur drive instead of the Fordson's worm. The Austin weighed 26 cwt compared with the Fordson's 23. Their horsepower ratings, when tested at St Germain en Laye in France in the spring of 1919, were 22 hp for the Fordson and 25 hp for the Austin.

The Fordson's engine – whether the Hercules or the Model T inspired unit that followed it in 1920 – had a bore and stroke of 4 by 5 inches for 4.1 litres (compared with 3.75 by 4 inches for 2.89 litres of the Model T) with a 3-bearing crankshaft, while the Austin's dimensions were 3.75 by 5 inches for 3.7 litres with an immensely robust crankshaft on five main bearings. Both claimed 2,500 lb drawbar tractive effort though, on test, there were wide discrepancies based on ground conditions as the Austin soon claimed 3,000 lb against the lighter Fordson's 1,800 lb. The Fordson began with 4-foot-diameter driving wheels but both makes soon quoted 42-inch wheels.

Price started off in the Fordson's favour and grew ever more divergent with the Austin reduced from £360 to £300 for 1922, when the Fordson cost a mere £120 – all of which helps to explain the Austin tractor's move to France.

As early as November 1919, Sir Herbert expressed the problem rather differently when he explained that 'the order book at Longbridge was full to overflowing and we have so many tractors to deliver that the output of the French works will be a great relief'. The estimated value of output in the 1920s was put at £9 million worth of cars, lorries and tractors, and investors were advised to buy 98,500 shares at 100 francs each to raise ten million francs (approximately £275,000) to ensure production at Liancourt from December 1919.

The big inducement was that 100 francs normally equalled £4 but at 'the present favourable rate of exchange is approximately only £2 12*s* 0*d*'. It seems that the share issue was not successful as the Austin Motor Co. ended up with around three quarters of them.

Sir Herbert went to France in March 1920, ostensibly to study cultivation between vines but more likely to look at Liancourt progress, which turned out to be very slow – only 200 tractors were made in the first year. A moulders strike at Birmingham a couple of months later saw the decimation of what was supposed to be 150 cars, 60 lorries, 60 tractors and 100 lighting sets per week. By June the problems had been resolved and 66 tractors were made in one week, with half sent abroad, though the broadly similar lorries (except that they had worm drive) were only then actually entering production. 1920 ended with few lorries but 4,319 cars built.

Austins figured prominently at the 1920 Royal Agricultural Show at Darlington and at the Highland Show, Aberdeen that summer. HM the King bought an Austin tractor for Balmoral in preference to the new 3x3 Glasgow tractor made in Scotland. Austins also figured in Chartres, Bourges, Jardin des Tuileries and other French trials.

In September, Austins performed well at the Lincoln Trials, though one ploughman was said to lack skill and three furrows was too arduous. Five tractors had replaced sixteen horses for Longbridge works transport, moving 100 to 120 tons per day. A good horse was said to cost £100 at the time and had upkeep of 4/6*d* daily. October saw the announcement that Longbridge would soon increase tractor output to 200 per week and 1921 started well with victories in three ploughing matches. Austin offered a £10 prize to farmers who could come up with the true cost of tractor versus horse ploughing in an effort to raise awareness of the land wasted on their fodder.

A problem with the tractor's aluminium pistons had been identified. These were fine when running on petrol in the cars and lorries but tended to break up when switched over to paraffin on the tractors. Harry Ferguson, who had kept all those US tractors running in Ireland and was by then an Austin dealer, helped come up with the solution of cast-iron pistons.

In February 1921, Sir Herbert wrote to his financial backers stating 'for the whole of the past year the business has been carried on under very great difficulties and restrictions in consequence of the shortage of capital, and the Directors have finally had to decide to make an issue of Debentures as the best and most economical means available at the present time to finance the increased trade of the Company'.

This appeal fell on deaf ears as on 26 April 1921 Sir Arthur Whinney was appointed as Receiver and Manager. Months of uncertainty followed but liquidation was successfully resisted and a new 12 hp economy car late that year went on to save the day in 1922 followed soon afterwards by the legendary Austin Seven. At last there was good news and the Midland Bank backed the restructure plans of Austin Finance Director, Ernest Peyton, while Carl Engelbach looked after production and Sir Herbert acted as Chairman.

Four Austin tractors were entered for Shrawardine Tractor Trials near Shrewsbury in September 1921. The ones running on paraffin cost 19.3 pence in fuel per acre compared with 40.6 pence for the petrol version. The Fordson managed 15.5 with Fiat and Case ahead of even that, but trailing the hot bulb oil-engined Swedish Avance on 5 pence. While the Austins did not shine in any particular area, they proved to be reliable and better than average among the twenty-seven entrants.

In tests the drawbar pull to skid wheels and the sustained pull on the best Austin was 2,000 lb and 1,820 lb compared with the Fordson's 1,750 lb and 1,385 lb respectively, and their drawbar horsepowers were 11.4 and 7.35 – figures that in some ways justified the much higher price of the Austin.

Henceforth the emphasis was on French production and it seems likely that the tractors finished at Longbridge in 1923, though unsold stock may have lasted into 1924 before French imports took care of what little demand there was. The British production has been estimated at 2,000 to 3,000. The price had fallen from £300 to £225 in 1923 and to £195 in 1924 before climbing back to £225 in 1925. It was the same in 1926, when there was also a £285 road tractor from France.

Sir Herbert Austin stepped down as a Conservative and Unionist MP in 1924, having found Parliament's protracted negotiations too tedious for a dynamic man used to getting his own way. He made the point that 'thriving conditions at Longbridge ... minimise the evils of unemployment [better] than could all my efforts on the floor of the House ... men outside such as myself can accomplish more outside than in'. That year he owned fifty shares in Ford of Canada, no doubt to keep an eye on the opposition.

His old firm of Wolseley had never recovered from the war and was sustaining heavy losses on its post-war vehicle production. Austin's major rival by then was Morris and the two men met in June 1924 to discuss the future with Dudley Docker of Vickers. Sir Herbert favoured a three-way merger, with William Morris in overall charge of the group, but Morris said he did not want to be answerable to anyone else. General Motors, whose Samson tractor had been a complete flop, looked at both Austin and Wolseley before settling on Vauxhall in December 1925, where its immensely successful Bedford commercial vehicles were made in the 1930s.

In 1927 Austin, freed at Longbridge of all but cars and vans, made 38,000 in his 62-acre factory with 8,000 workers.

In February 1927 the bankrupt Wolseley came up for sale and an unnamed American firm and Sir Herbert were under bidders to Morris (who would become Sir William in 1929). He paid £730,000 which was rather less than the firm owed but it released Vickers, who had found great difficulty in replacing military with civilian orders and would merge with Armstrong-Whitworth in 1927. It had tried car bodies, sewing machines, cheaper Stellite cars and even tractors at its Maxim works in Crayford, Kent. The mid-1920s to early 1930s tractors were based on the International Harvester 15/30 and were intended for Britain's colonies and dominions where, unlike the US products, they would benefit from colonial tariffs. Known for a time as Vickers-Aussie, they were not a commercial success.

Armstrong-Vickers also made small Ford-engined crawler tractors for agricultural purposes but it was their military role, notably as Carden-Lloyd Bren gun carriers, where their true significance lay. Later based on Second World War battle tank technology, Vickers-Armstrong would build big Rolls-Royce powered crawlers, mostly for earthmoving purposes, from the 1940s to early 1960s.

As we have seen, the early motor industry was remarkably incestuous, with seemingly everyone knowing everyone else and having various levels of financial or technical involvement. It is perhaps worth mentioning that John Davenport Siddeley, who took over from Herbert Austin at Wolseley, went on to found his own car firm that joined Armstrong-Whitworth in 1919 and made Armstrong-Siddeley cars. Around 1930 it also tried tractors, this time built under licence from Pavesi of Italy. These were advanced in featuring 4x4 with centre-pivot steering, but they fared far worse commercially than the Italian originals. By strange coincidence another Italian business, the Terni Shipyards, had been linked with Vickers, Sons & Maxim since 1900 and had become a major arms supplier before turning to farm tractors. As Odero Terni Orlando (OTO) it rose to fourth position in Italian production in the 1950s.

Austin had plainly made the right decision in concentrating Longbridge on the vehicles that sold best as the tractor market became extremely crowded in the mid-1920s (see table of makes and specifications on page 57). Around 1930 it would become even more competitive when London bus and lorry maker AEC introduced its General (soon renamed Rushton), Lanz from Germany appointed a British importer, Massey-Harris set up shop near Ford's old Trafford Park factory, and Marshall, McLaren and Garrett brought in pioneering British diesel tractors.

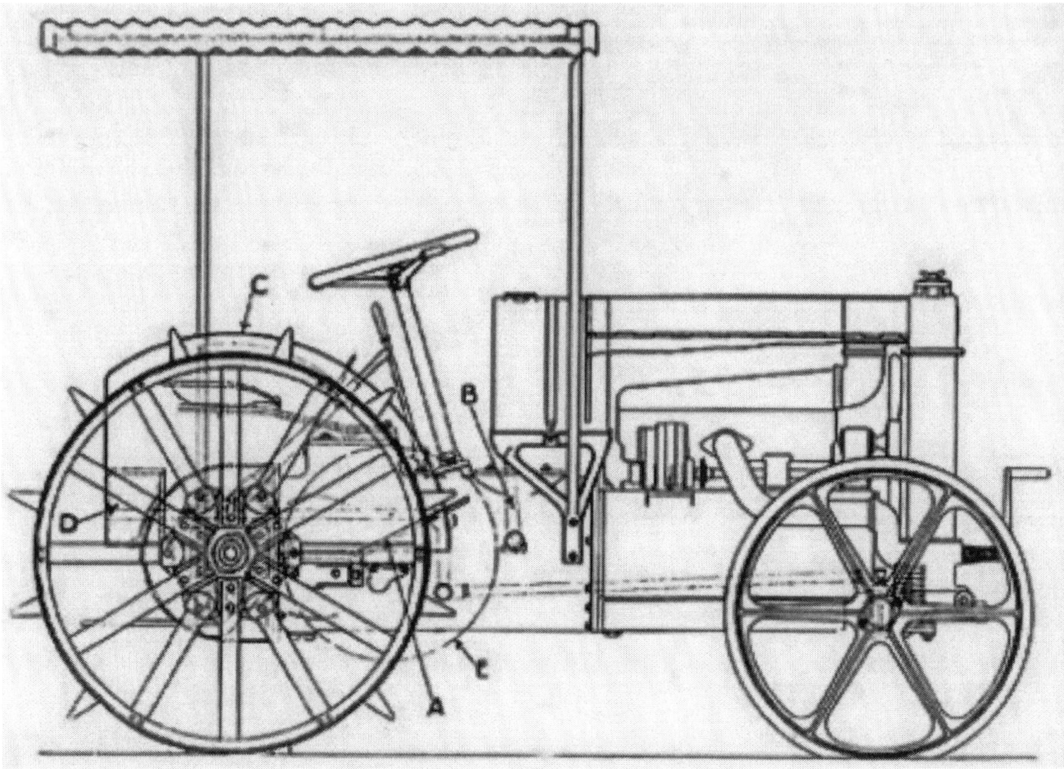
C
B
D
E
A

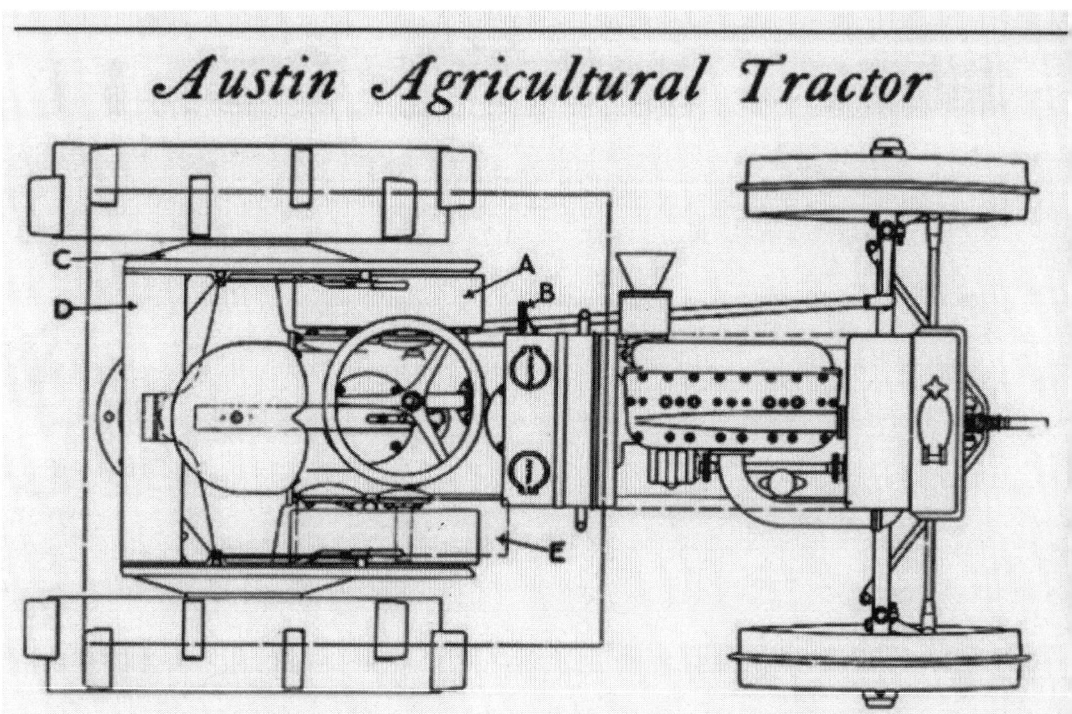
Austin Agricultural Tractor
C
A
B
D
E

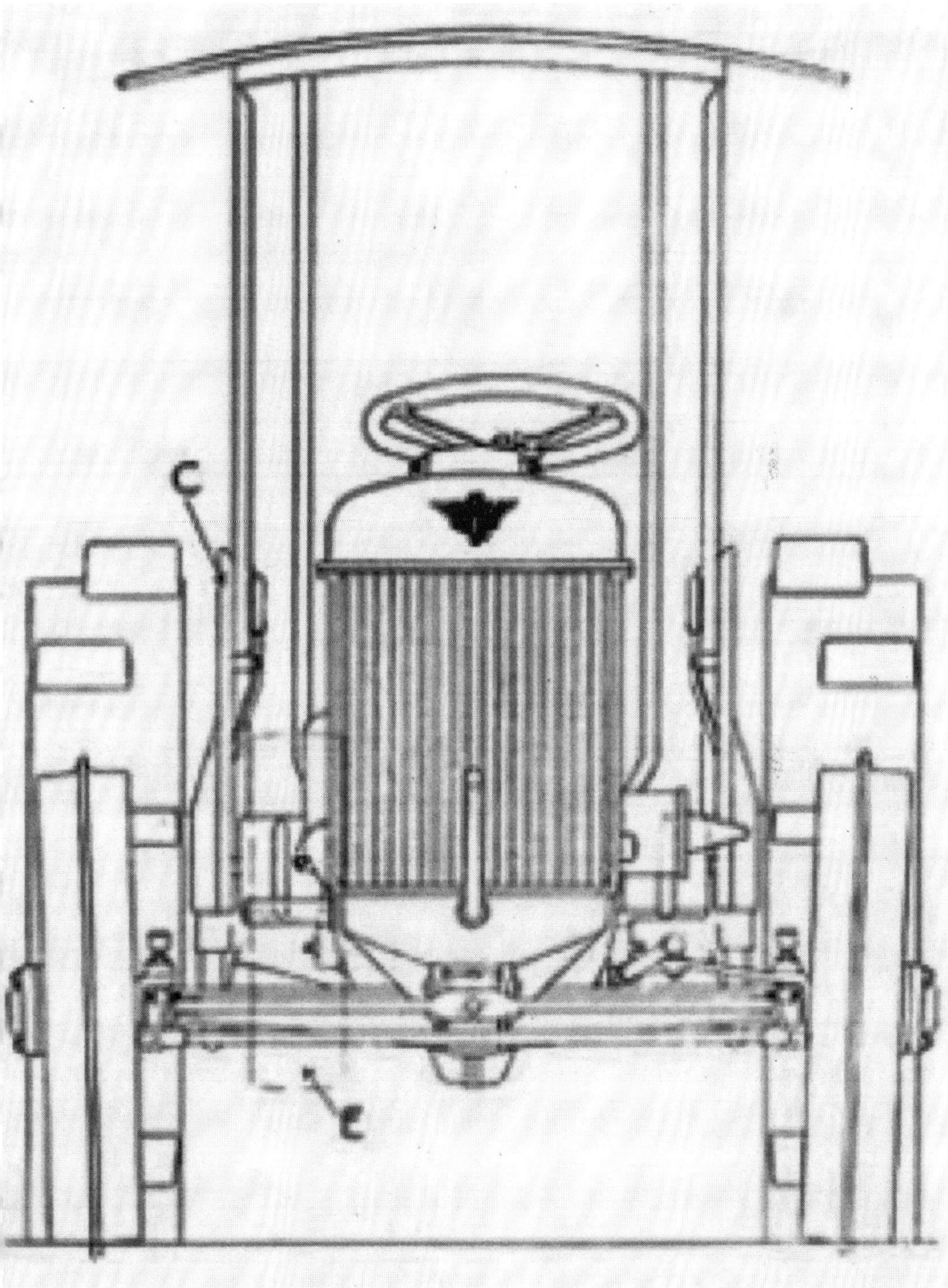

Opposite page and above: The first outline views of the new Austin from the side, above and front and an early prototype being driven in allegedly 1918 is seen on page 40.

This artist's impression of the Austin made it look particularly compact due to the large size of the driver.

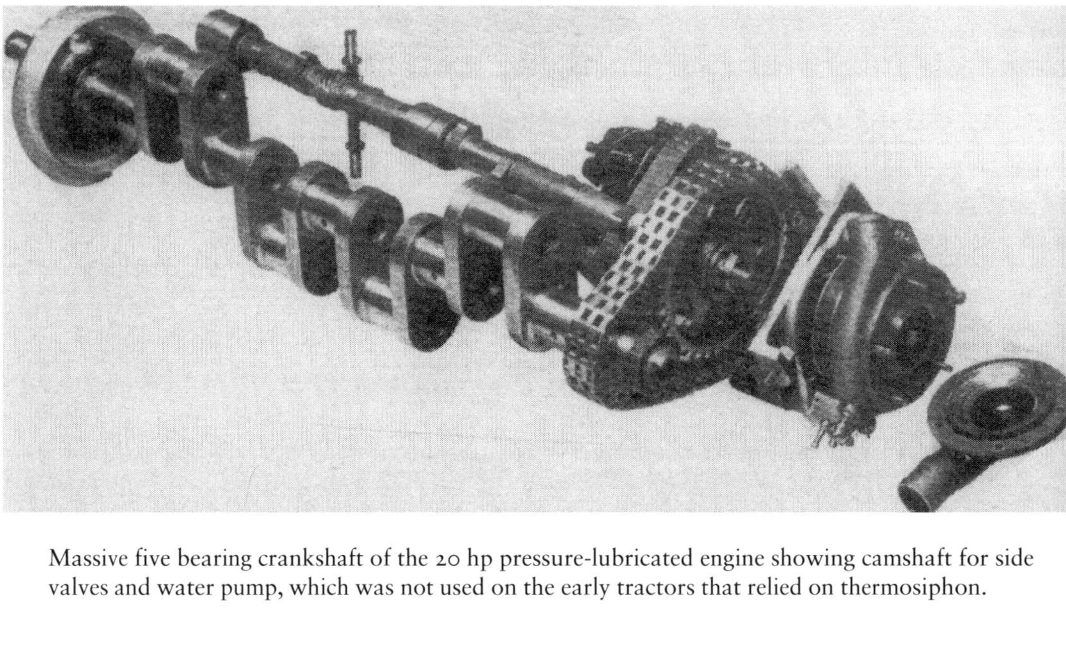

Massive five bearing crankshaft of the 20 hp pressure-lubricated engine showing camshaft for side valves and water pump, which was not used on the early tractors that relied on thermosiphon.

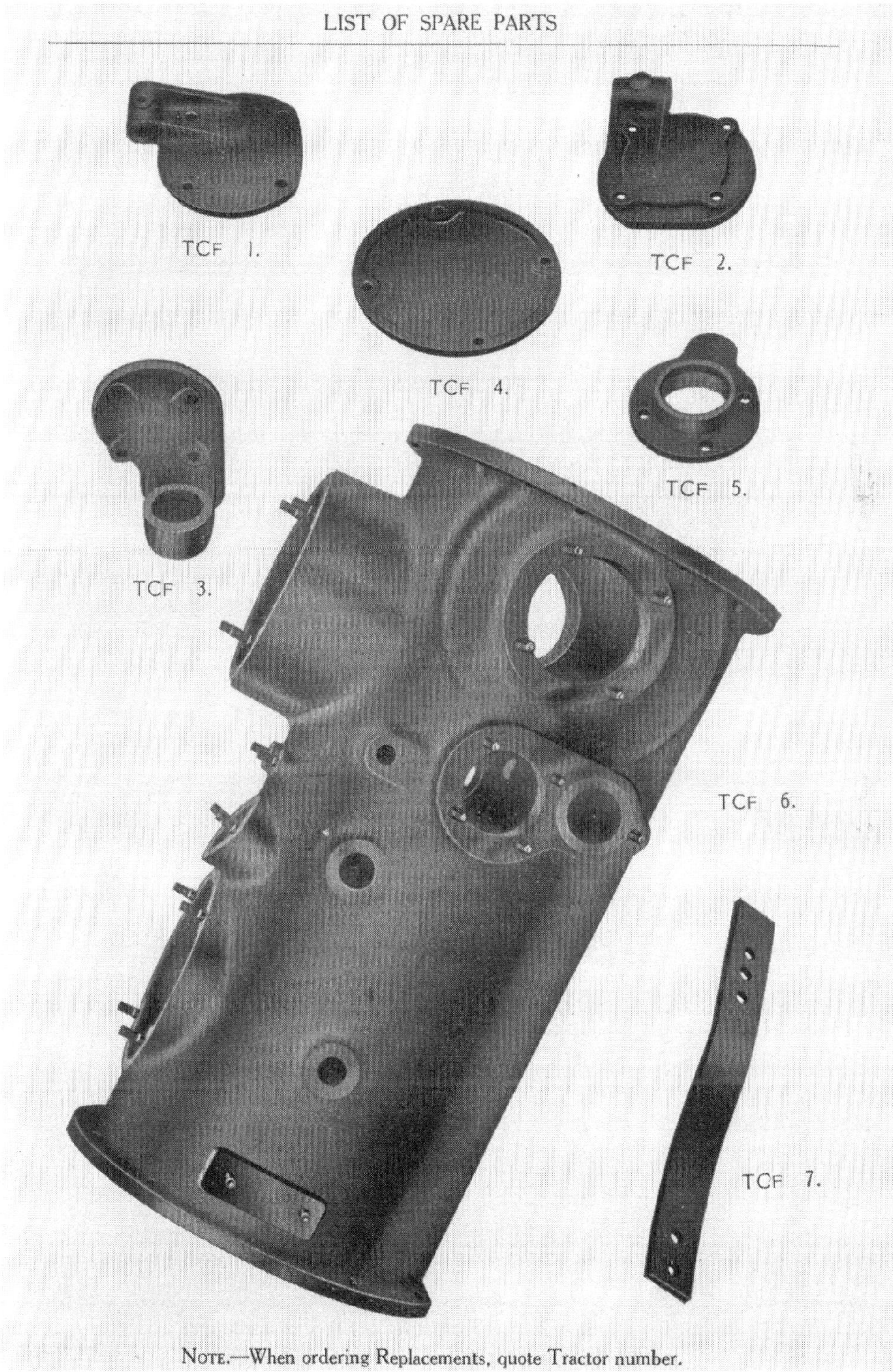

To cope with precision engineering demanded for armaments in the 1914–18 war, Austin had invested in a metallurgical laboratory, which stood it in good stead for the strongest components made from the minimal metal on its vehicles. This is a cast tractor transmission casing.

THE AUSTIN FARM TRACTOR.
THE ENGINE.
The Austin Farm Tractor with bonnet removed, showing engine. The off-side view above shows plainly the thermo-syphon cooling circuit, the air cleanser, the magneto and its drive, also the crankcase oil filler.
The top view shows the road hauling pads fitted on wheels ; the bottom view shows the tractor fitted with spuds on the driving wheels and angle irons on the front ones, for field work.
The view above shows the carburetter, and its air supply passing through the cleanser and exhaust-heated box. Valve stems, springs and adjustable tappets can be seen behind the carburetter, the two-way petrol and paraffin tap under the tank, and the clutch pedal.

Note.—When ordering Replacements, quote Tractor number.

Above: The first but undated Austin Tractor List of Spare Parts featured these views and did not refer to a water pump, though the part number for a radiator fin was TEM 9, of which about 9,000 were required.

Left: When restoring Austin tractors there is uncertainty about the exact size of the lettering. Part number TEO 1 shows that it really was written large.

TRACTORS TO SUIT ALL
WORKING CONDITIONS.
"THE BLACKSTONE"
BLACKSTONE CO., LTD., STAMFORD.
BRONZE
MEDAL
1920.
"THE PETERBRO"
PETER BROTHERHOOD, Ltd., PETERBOROUGH
Manufactured by ASSOCIATED FIRMS OF
AGRICULTURAL & GENERAL ENGINEERS, LTD.,
CENTRAL HOUSE, KINGSWAY, LONDON, W.C.2.

Agricultural and General Engineers tried to be the British equivalent of International Harvester but without much success. This advertisement dates from 1921 and shows two models that outlasted the Austin in British production.

A POSTCARD
will bring by return of post a copy of the booklet entitled "New Century Farming," which tells the story in detail of the
Austin
AGRICULTURAL TRACTOR.
This tractor is designed to be of the utmost use to the British farmer. It is low in price, economical on fuel and is capable of doing every type of farm work both mobile and stationary. EVERY SILVER MEDAL presented in England in 1919 by Agricultural Societies for Ploughing was won by the Austin.
The AUSTIN MOTOR Co., Ltd.,
NORTHFIELD - BIRMINGHAM.
Austin

Postcard sent to farmers in 1920 mentioning silver medal (but no gold) success in 1919.

Some of the five tractors used for Austin works transport at Longbridge in 1920, when they replaced sixteen horses.

1919 tractor with canopy showing off its maximum front axle tilt from its unsprung central mounting point. It had two forward gears, though a third became available that autumn.

Above: Longbridge works view of a steam locomotive collecting a large consignment of tractors.

Left: Loading crated tractors by railway steam crane at the Longbridge branch line.

Above: Foster of Lincoln threshing machine powered from an Austin's belt pulley. Paraffin consumption was quoted at 1.5 gallons per hour.

Right: November 1919 letter from H. Austin, with no mention of his knighthood, in which he outlined the opportunity for investors.

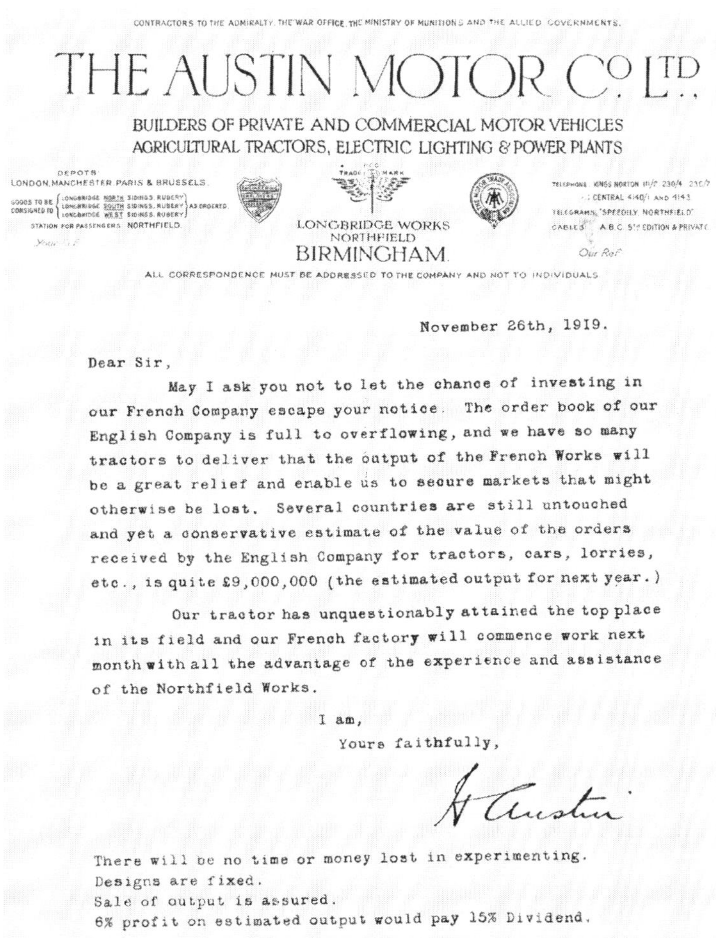

CONTRACTORS TO THE ADMIRALTY, THE WAR OFFICE, THE MINISTRY OF MUNITIONS AND THE ALLIED GOVERNMENTS.

THE AUSTIN MOTOR C^O. L^TD.

BUILDERS OF PRIVATE AND COMMERCIAL MOTOR VEHICLES
AGRICULTURAL TRACTORS, ELECTRIC LIGHTING & POWER PLANTS

DEPOTS
LONDON, MANCHESTER, PARIS & BRUSSELS.
GOODS TO BE CONSIGNED TO { LONGBRIDGE NORTH SIDINGS, RUBERY / LONGBRIDGE SOUTH SIDINGS, RUBERY / LONGBRIDGE WEST SIDINGS, RUBERY } AS ORDERED.
STATION FOR PASSENGERS: NORTHFIELD.

TRADE MARK

LONGBRIDGE WORKS
NORTHFIELD
BIRMINGHAM.

TELEPHONE: KINGS NORTON 111/2, 230/4, 230/7
CENTRAL 4140/1 AND 4143
TELEGRAMS: "SPEEDILY, NORTHFIELD."
CABLES: A.B.C. 5TH EDITION & PRIVATE.

Your Ref. Our Ref.

ALL CORRESPONDENCE MUST BE ADDRESSED TO THE COMPANY AND NOT TO INDIVIDUALS

November 26th, 1919.

Dear Sir,

May I ask you not to let the chance of investing in our French Company escape your notice. The order book of our English Company is full to overflowing, and we have so many tractors to deliver that the output of the French Works will be a great relief and enable us to secure markets that might otherwise be lost. Several countries are still untouched and yet a conservative estimate of the value of the orders received by the English Company for tractors, cars, lorries, etc., is quite £9,000,000 (the estimated output for next year.)

Our tractor has unquestionably attained the top place in its field and our French factory will commence work next month with all the advantage of the experience and assistance of the Northfield Works.

I am,
Yours faithfully,

H Austin

There will be no time or money lost in experimenting.
Designs are fixed.
Sale of output is assured.
6% profit on estimated output would pay 15% Dividend.

The author on the left, next to owner Tim Pearce, lines up R815 for a photograph. Though not road registered until 1921, 815 suggests 1919 production.

The very original green R815 with red wheels and yellow lettering shown in the barn where it had stood since 1952. The significance of the name Iron Duke is unknown. Some early Austins appear to have been dark blue.

Right: Another early preserved example is R2260 in Eire, which has 27/10/1920 on its transmission casings.

Below: A much-modified Austin at Jondaryan Woolshed, Darling Downs, Queensland, Australia. Only the early ones had right-angle edges to their radiator cores. Behind is one of GMC's unsuccessful Samsons along with a Fordson and a Case.

July 1920 advertisement trumpeting 200 vehicles produced in a week and showing 20 hp cars and tractors.

A clearer view of the tractors in the July 1920 advertisement in the yard of South Works. They do not quite reach the sixty-six quoted in the record week.

The author's late father-in-law is shown as a boy on the reaper binder, hauled by this early Austin at work in Oxfordshire in the 1930s.

A 1924 photograph of a family group and Austin at Merrytown Farm, Lanarkshire, part of the Duke of Hamilton's estate.

Austin with square radiator core, this time preserved in Somerset. At least another dozen are known to exist of the British variety.

Tractors Nos. 1, 2, 3 & 4 — Austin

The Austin tractor is manufactured in England at Northfields, Birmingham, by the Austin Motor Co., Ltd.

In general construction this tractor is of the unit (frameless) type. The crankcase of the engine bolts rigidly to the gear box and the gear box in turn bolts rigidly to the back axle casing, thus forming one solid unit.

The radiator is of usual tubular type and is cooled by means of a fan directly behind it. The water circulation is maintained by means of a centrifugal pump. The engine has four cylinders; lubrication is by pressure from a gear-driven pump through a hollow crankshaft, which ensures direct service of oil to all crankshaft and connecting rod bearings.

The gear box is of the sliding pinion type and provides two speeds forward and one reverse. The gear box also embodies a cross shaft for belt pulley, the size of which is 24 in. dia. by 6 in. wide; it runs at 360 revolutions per minute at normal engine speed. Final drive is by parallel faced spur and crown wheels.

The front of the machine is sprung and pivot mounted on the axle by Austin patent construction.

Steering is by worm gearing (totally enclosed in the gear box) and Ackerman front wheel mounting.

The Austin farm tractors using oil (paraffin) and spirit fuels (petrol and benzol) differ in regard to:

(1) Compression ratios,
(2) Induction pipes,
(3) Degree and means of heating the charges before admission to cylinder.

Specification of this machine in tabular form will be found in Tables I and II on pages 31 to 39.

This is one of the Austins at the tractor trials at Shrawardine near Shrewsbury in September 1921. It featured a water pump, suggesting that earlier thermosiphon types overheated. A stronger back axle and differential plus thirty per cent more drawbar hp than previously were claimed.

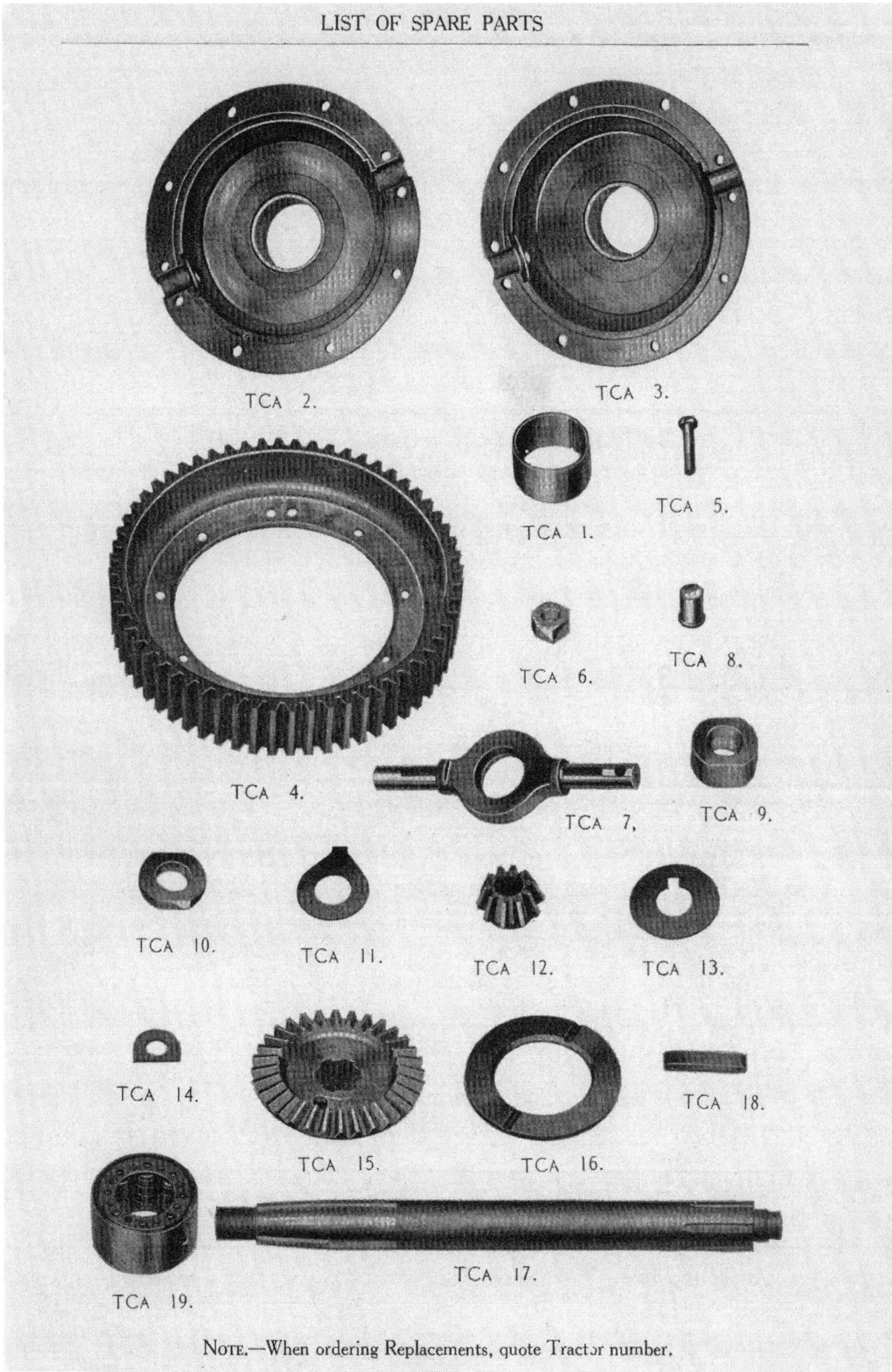

What the rear axle internals looked like in 1920.

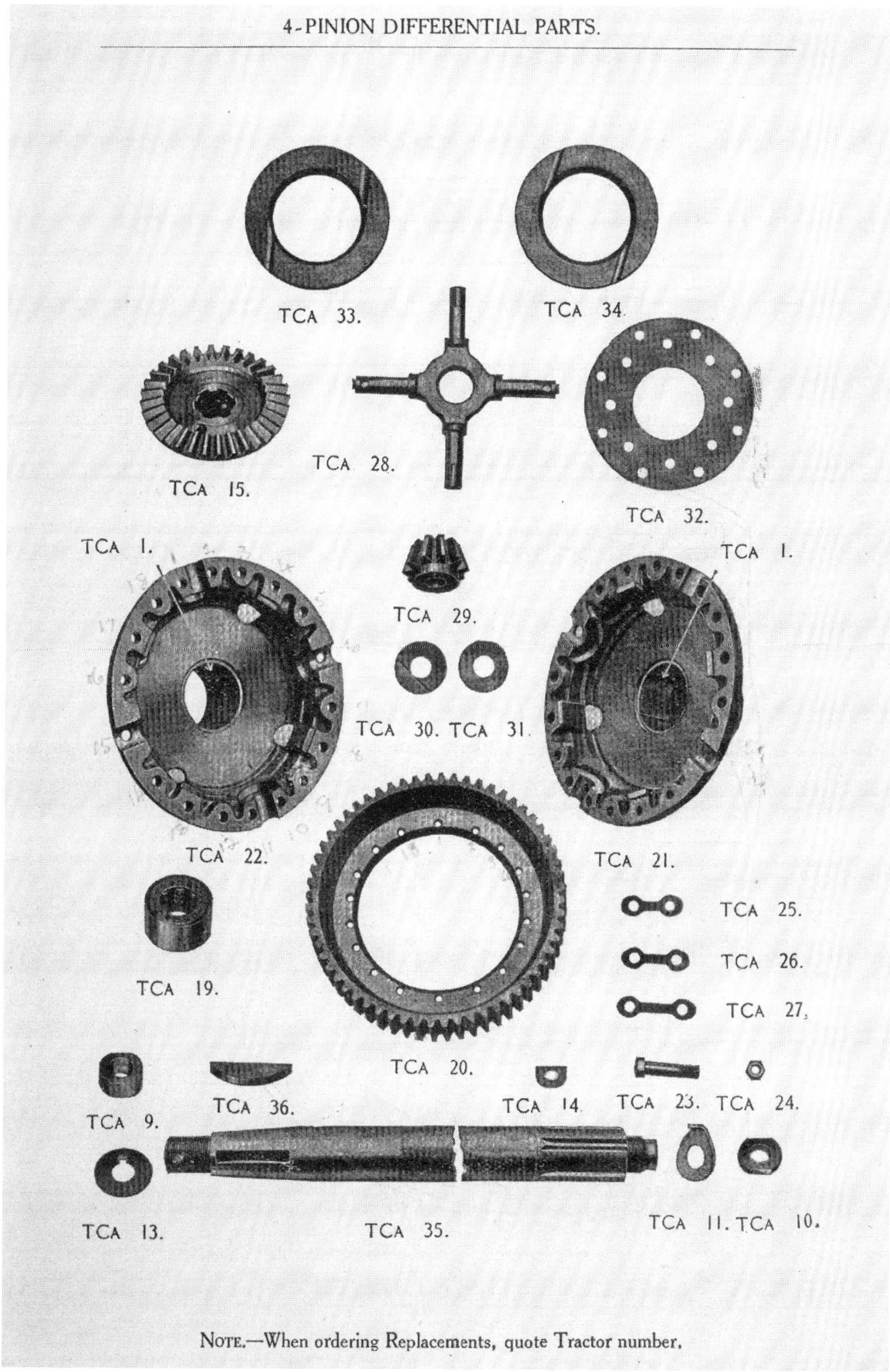

This diagram shows the beefier rear axle internals and is dated August 1921.

This is Austin's revised post-war 20 hp 1.5-ton capacity lorry. The radiator and bonnet looked just like the tractor, though the exterior steering column and axles differed.

After a twenty year absence since Herbert Austin's Wolseley tractor, Vickers unwisely returned to the fray with this official copy of an International 15/30, primarily for Australia.

AGRIMOTORS—BRITISH AND FOREIGN.

Name of Agrimotor.	Maker or Concessionnaire.	Reference No.	Model.	Engine: Rated Horse-power.	Engine: No. of Cylinders.	Engine: Bore and Stroke.	Final Drive.	No. of Forward Speeds.	Belt Drive: Pulley Diam.	Belt Drive: Pulley Width.	Belt Drive: Revs. per min.	No. of Brakes.	Springs	Front Wheels: No.	Front Wheels: Diam.	Front Wheels: Width.	Rear Wheels: No.	Rear Wheels: Diam.	Rear Wheels: Width.	Wheelbase.
AUSTIN	Austin Motor Co., Ltd. ...	A 1	—	25	4	3¾″ × 5″	S.G.	2	24″	5″	360	2	F.	2	30″	6″	2	42″	10″	5′ 8″
BLACKSTONE	Blackstone & Co., Ltd. ...	2	—	30	3	5½″ × 6½″	I.G.	3	—	6″	505	2	—	2	—	—	—	—	—	—
CASE	Associated Manufacturers' Co. (London), Ltd.	3	12–20	27	4	4⅛″ × 5″	S.G.	2	14½″	6⅜″	1050	1 or 2	Nil	2	30″	6″	2	42″	12″	5′ 5″
		4	15–27	35	4	4½″ × 6″	S.G.	2	16″	7½″	900	1 or 2	Nil	2	32″	6″	2	52″	12″	6′ 4½″
CLETRAC	H. G. Burford & Co., Ltd....	5	Chain track	25	4	4″ × 5½″	I.G.	1	8″	6″	—	—	—	—	—	—	—	—	—	—
CRAWLEY	Crawley Agrimotor Co., Ltd.	6	—	30	4	4⅛″ × 5½″	G.	2	14″	6″	500 or 625	1	Nil	2	48″	8″	1	17″	2¾″	—
FIAT	Fiat (England), Ltd. ...	7	702A.	18–25	4	mm. 105 × 180	W.	3	13″	6½″	100 to 750	1	F.	2	32″	5″	2	52″	12″	5′ 9″
FORDSON	Ford Motor Co. (England), Ltd.	8	—	25.6	4	4″ × 5″	W.	3	9½″	6½″	1000	1	Nil	2	27″	5¼″	2	42″	12″	5′ 3″
FOWLER	John Fowler & Co. (Leeds), Ltd.	9	—	20	4	3¼″ × 4½″	G.	2	6″	4½″	1200	—	—	2	44″	6″	1	12″	4½″	8′ 0″
		10	—	30	4	4″ × 5¼″	G.	2	8″	4¾″	1000	1	—	2	54″	8″	1	18″	5″	10′ 0″
		11	—	40	4	4½″ × 6¼″	G.	3	12″	6½″	1000	1	—	2	63″	8″	1	19½″	5″	13′ 6″
HART-PARR	British Hart-Parr Co. ...	12	30	28.8	2	6½″ × 7″	I.G.	2	14″	8″	750	2	Nil	2	28″	5″	2	52″	10″	7′ 5″
		13	20	18.6	2	5¼″ × 6½″	I.G.	2	14″	8″	750	2	Nil	2	28″	5″	2	46″	10″	5′ 6″
MANN	Mann's Patent Steam Cart & Wagon Co., Ltd.	14	—	22	2	4″ × 8″	G.	3	30″	6″	300	2	—	2	35″	8″	2	51″	20″	8′ 0″
NEW INTERNATIONAL	International Harvester Co. of Great Britain, Ltd.	15	Junior	28.9	4	4¼″ × 5″	S.G.	3	15¼″	7″	645	1	—	2	30″	4½″	2	42″	12″	6′ 6″
		16	Heavy duty	32.5	4	4½″ × 6″	S.G.	3	16¾″	8″	595	1	—	2	34″	6″	2	50″	12″	7′ 1″
PARRETT	Agri-Tractor Contract Co., Ltd.	17	—	25.8	4	4¼″ × 5½″	I.G.	3	12″	7½″	1000	1	Nil	2	46″	4″	2	60″	10″	7′ 10″
PETERBRO'	Peter Brotherhood, Ltd. ...	18	—	30–35	4	4¾″ × 5½″	I.G.	2	12″	6″	900	1	—	2	36″	6″	2	54″	10″	7′ 3″
RENAULT	Renault, Ltd.	19	—	24.8	4	mm. 100 × 160	D.R.	3	8″	6″	1400	1	—	2	27½″	5½″	2	46″	12″	5′ 10″
		20	Chain track	24.8	4	100 × 160	D.R.	3	8″	6″	1100	1	—	—	—	—	—	—	—	—
SAUNDERSON	Saunderson Tractor & Implement Co., Ltd.	21	—	23.6	2	5½″ × 8″	I.G.	3	12″	7″	750	2	F.	2	30″	6″	2	48″	10″	7′ 6″
SIMAR	Piccard Pictet & Co. (London), Ltd.	22	2-wheeler	8–10	1	mm. 95 × 90	B.	2	—	—	—	—	—	2	18″	5″	—	—	—	—
WALLIS	Ruston & Hornsby, Ltd.	23	—	28	4	4¼″ × 5¾″	G.	2	18″	6¾″	430	1	—	2	30″	8″	2	48″	12″	7′ 0″

LIST OF ABBREVIATIONS.—FINAL DRIVE: B. bevel; D.R. double reduction; G. gear; I.G. internal gear; S.G. spur gear; W. worm. SPRINGS: F. front.

An idea of the competition Austin faced in 1925. Plainly these are the UK and foreign tractors available in Britain, hence no mention of Vickers.

Chapter 4

French Adventure

The French factory at Liancourt was described in a document assigning its mortgage to the Austin Motor Co. by A. Kauffmann (a director in 1924 and still there in 1930) to Mr Ronsseray (presumably at Longbridge) as a large brick-built space surmounted by a clock reached from the Rue Victor Hugo past the door keeper's lodgings, a little garden and a yard leading into workrooms and offices. A large store building adjoined to the left by a little kiosk of stone and brick housing a water closet. Behind were the buildings and outhouses plus a big chimney (what purpose it had served was not explained). The size of any of this was not referred to but the site seems to have been the same 1,300-metre length as Bajac next door, separated by a path with a slaughterhouse at its end dating from the days of leather production on the site. Opposite were more buildings with a sawmill powered by the River Beronnelle and another yard and buildings used by a wine merchant and landlord of the Buvette de la Cascade bar.

In addition to a small stately home in the Duke de la Rochfoucauld's botanic garden bought by Austin for their managing director, 325 acres of farmland was acquired for expansion and agricultural machinery tests. Included with the property was standing timber worth £5,000.

In 1919 Sir Herbert Austin had stated that the works were needed to free up capacity at Longbridge for its full order book. However, the share flotation documents for the French business spoke of greatly increased restrictions on imports into France but that a demand for tractors existed there, for which the Liancourt works would be able to build 2,000 a year. Longbridge had supplied Liancourt with a power plant, machines, small tools, gauges, jigs, fixtures and plans to ensure full production by the end of June 1920. On this basis, dividends in excess of ten per cent could be anticipated, though, as noted in Chapter 3, the take-up was poor and the Austin Motor Co. or Sir Herbert ended up with three quarters of them.

To start with, the tractors were identical to the British counterparts, which had opened up the French market. Even their threads were not metric, which soon caused problems for dealers, repairers and farmers.

Directors from the Austin Motor Co. were Sir Herbert, who had ten shares, and Randle G. Ash joined by Alfred Tebbitt of the British Chamber of Commerce in Paris, and A. Gest, the managing director of agricultural machinery dealers Maison Th. Pilter. Despite import tariffs, the Fordson sold by Maleville et Pigeon of Chartres had accounted for 1,000 sales by the time of the Chartres Motoculture Trials of October 1920. Maybe because of difficulties with selling Austins, Maison Pilter refused to take any more (soon after advertising Austin in the Chartres programme) and thus lost their exclusive five-year distribution contract, leaving SA Austin to find new dealers. Pilter went on to sell Bauche tractors and, later in the 1920s, gained the Case franchise.

Surviving Austin R number 1336 is said to have been made in France in 1921, proving that grandiose plans for the 2,000 per year were optimistic. It also begs the question of whether it really was French or one of the British batch stamped with similar numbers. On 1 January 1920 an agreement had been signed between the two Austin firms for Liancourt to use the same name and patents and to pay royalties to Longbridge of one per cent, and 1.5 per cent if profits were sufficient to pay six per cent dividend to shareholders. The fees and travel expenses of visiting British experts would have to be paid and Longbridge wouldn't sell tractors to France or its dependencies (nor France to Britain without prior agreement).

Another SA Austin director of the early 1920s was stated to be H. G. Burford, who had been involved with numerous makes of vehicle since 1900 and had more recently acquired British rights to the Cletrac crawler tractor from America. He became President of the Institute of Automobile Engineers in 1923, having given lectures about tractors to its members.

In 1923 the Austin Motor Co. loaned SA Austin £15,000 at six per cent interest in exchange for all its assets (which is presumably when director Kauffmann handed over the mortgage). There were five years to repay and SA Austin promised also to pay unstated royalties from the previous two years plus a sum from the sale of workmen's houses (which still hadn't been settled in 1924, suggesting cash flow difficulties).

From December 1923 to June 1924 a visitor from Longbridge charged £137 *6s 5d* for his services. This may have been French speaker George Coates, who had been apprenticed at Longbridge and who is known to have been sent to Liancourt in 1924 'to investigate a manager who had hopped off with £40,000 and some of the drawings'. An unverified production figure of 1,500 was quoted for that year but royalty payments for January and February covered a mere twenty tractors. A dealer later recalled 10 per month in the early days leading to 20 or 30 per month after two or three years, then 80 per month followed by 130 and easily 150 in 1930.

These royalties changed to five per cent on net cash received on sales, with 2.5 per cent on Turkish sales, and director Kauffmann referred to a debt of £10,000 to Austin Motor Co. in addition to the £15,000 borrowings. Sir Herbert or his companies started to build up a big shareholding in Acieries du Mons (probably the source of castings) and in 1926 Adrien Roussel, who had been a machinist at Liancourt since 1922, became Works Manager of SA Austin – a position he held to 1936 and further promotion. A young contemporary of his was Gabriel Naninck who had joined at the age of fourteen in 1921 and worked his way up to Technical Director in the 1930s.

There were no Austin tractors at the Royal Agricultural Show held at Reading in 1926 and in France a new DE model became available as an IBA solid rubber-tyred road tractor. The DE may well have been a metric version of the old R and the Payen Gasket Guide of 1938 gives 1925 as the year of the arrival of a different cylinder head gasket. The DE26 had this same head gasket and featured two detachable sections that could carry road bands or cleats on the drive wheels. In addition there was also an ultra-narrow V model, for vineyard use.

The year of 1927 produced a French profit of £948 0*s* 1*d*, to which was added £2,112 18*s* 0*d* brought forward. Sir Herbert was owed £2,800 at the time and the Austin Motor Co. a further £2,736 5*s* 11*d*. There had been close ties with the neighbouring Bajac implement firm and experiments took place with mounted ploughs and mechanical lifts.

An electrical fire that engulfed a fifty by eight-metre building at SA Austin in 1928 destroyed components as well as the design office and its plans and documents. Two million francs was the estimated damage, though from this misfortune, which was fought by several brigades including Albaret's (*see Chapter 2*) own firemen, one learns that 400 men were employed on the site. Production soon returned to normal and by early 1929 over 4,000 Austin tractors were said to be at work – a figure that had reached 5,000 in September. The latest ones were 15/25 hp models called B028, which became the three-speed SA3 and V33 with governors and air cleaners in 1929. The B028 had improvements to the clutch and gearbox and a declutchable pulley and PTO shaft.

Of the 822 tractors of all sorts imported into Britain in 1928, ninety-six per cent came from the USA and presumably none or very few from SA Austin. Sir Herbert was fully vindicated in having withdrawn from the UK market when Britain's tractor exports achieved a miserable eighty-six in 1928.

As of July 1928, all royalties had been paid up to the end of 1926 and Sir Herbert bought all further provisional royalty payments due to the Austin Motor Co. for £1,000 – it would be interesting to know if he ever got his money back.

He was expanding worldwide with over a quarter of Longbridge production exported and new opportunities opening up for the legendary Austin Seven car in France (where Rosengart, rather than the Liancourt factory, started licenced production), in the USA with American Austin, and in Germany where the Dixi licence soon passed to BMW.

1930 marked the arrival of cheaper and more basic Simplyx models to counter sales resistance and the return of SA Austin to the British market with an appearance at the Royal Show, as well as the World Tractor Trials. Importation was handled by R&J Park Ltd of Dominion House, Thames Road, Chiswick, London, W4, who referred to the French Austin's Headquarters at 139 Rue Lafayette, Paris, rather than the factory address. A Monsieur Bureau (conveniently meaning 'office') worked there and Sir Herbert had reason to write to him enquiring where some of his share certificates were – the only known survivor refers to 1,425 at 100 francs each in 1921. The price of the 3¾ bore by 5 inch stroke 15/25 tractors weighing 4,060 lbs was now only £210 due to favourable exchange rates. One Austin ran on paraffin at Wallingford and one on petrol, and both were rated at fair to good. The petrol one had governor and carburettor problems while the paraffin one was adjusted for cruder fuel than was available. Their compression ratios were a lowly 4 and 3.5 to 1 respectively.

In 1931 Sir Herbert threw his weight behind the tractors, of which he was described as the designer, and R&J Park became known as the London branch of Société Anonyme Austin. They had a prominent display at the Royal Agricultural Show at Warwick, some twenty miles from Longbridge, and included a new sidevalve model (soon redesigned with overhead valves) with 4⅛ by 5 inch bore and stroke for a capacity of 4.4 against the original 3.7 litres, and 4.2 to 1 compression that developed 35 bhp at the belt and over 20 at the drawbar.

In the year to end July 1931 the Austin Motor Co. declared £1.078 million in profit, which, despite the aftermath of the Wall Street Crash and gathering Depression, had only decreased by £300,000 a year later. No figures were quoted for Liancourt.

There were nine Austin tractors at the Paris farm machinery fair in January 1932, including a beautifully sectioned one – possibly the same that had appeared there in 1930. At the fair a Memini carburettor for switching Austins and Fordsons gradually from petrol to paraffin was displayed, but in 1932 these became old hat when SA Austin launched diesel versions.

TRACTEUR “ AUSTIN ”

Le caractère principal de ce nouveau tracteur est sa grande légèreté alliée à une grande puissance. Il est fort maniable et peu encombrant. Le châssis est remplacé par un bloc compact formé de la réunion de trois carters en fonte aciérée renfermant tout le mécanisme qui de ce fait se trouve complètement à l'abri de la terre, de l'eau et de la poussière.

Dans un avenir prochain ce tracteur sera construit en France.

PRINCIPALES CARACTÉRISTIQUES

Puissance du moteur fonctionnant à l'essence		25 HP
Effort sur la barre de traction		20 HP
Nombre de cylindres		4
Alésage et course	millimètres	95 × 127
Nombre de tours par minute		1200
Diamètre et largeur de la poulie	millimètres	570 × 150
Nombre de tours de la poulie		330
Capacité du réservoir à pétrole	litres	45
— du réservoir à essence	—	5
Vitesses avant 1re	à l'heure	4 kilom.
Vitesses avant 2e	—	7 —
— arrière	—	3 — 250

Roues avant	Diamètre	762 mill.
—	Largeur	152 —
Roues arrière	Diamètre	1067 —
—	Largeur	254 —

Encombrement :

Largeur extrême		1550 mill.
Longueur —		2796 —
Hauteur —		1400 —
Empattement		1727 —
Cercle de virage		6 mètres 70
Poids en ordre de marche	kilogrammes	1450

MOTEUR monobloc à culasse amovible fonctionnant à l'essence, au pétrole, au benzol ou à l'alcool. Arbre vilebrequin supporté par cinq paliers. Régulateur centrifuge complètement enfermé.

REFROIDISSEMENT par thermo-siphon, le radiateur de grandes dimensions étant placé en avant du moteur et refroidi par un grand ventilateur commandé par engrenages enfermés.

CARBURATEUR ZENITH et vaporisateur chauffé par les gaz d'échappement assurant la complète carburation du pétrole.

ALLUMAGE par magnéto à haute tension.

GRAISSAGE DU MOTEUR par pompe sous pression.

EMBRAYAGE très sûr consistant en un cône d'acier s'engageant sur des segments amovibles garnis de Férodo.

TRANSMISSION par engrenages en acier cémenté baignant dans l'huile. Les arbres sont montés sur paliers à rouleaux.

PRIX SUR DEMANDE

Pilter's prospectus for an early tractor, quite possibly British-built.

March 1920 in the Paris Jardin des Tuileries and several tractors including (above) Fordson and this new Austin (below) are tested.

VIEW OF THE LIANCOURT WORKS. AUTOMOBILES ANGLAISES AUSTIN STE. AN.

THE Works are situated about half-a-mile from the Station of Liancourt, on the main line from Calais to Paris, four miles from the important industrial town of Creil, and about 35 miles from Paris. Facilities exist for siding accommodation.

The freehold property includes, besides the Works—as shown on the above photograph—two excellent houses and grounds for the Technical and Commercial Managers. On the opposite side of the main road, facing the Works, and included in the purchase, is the Park and farm lands, extending to nearly 300 acres, of the Duc de la Rochefoucauld, with the old chateau and extensive outbuildings, a saw mill, carpenter's shops and several dwelling houses. On the property it is estimated there is about £5,000 worth of timber, which it is proposed to cut down and dispose of. The farm lands will enable the Company to continuously test and keep the tractors up to date, train the employees of clients in actual farm work, and bring in a good return on the produce grown. The river Beronnel runs through the property, and provides all the water required for the Works.

After several uses and changes of ownership, the Liancourt factory was named Usines ('works') Austin.

QUELQUES VUES DE NOS USINES
Les Tours, Le Fraisage, La Centrale.

Three views inside the works showing powerhouse that supplied electric motors operating line shafts in the machine shops.

View of a machinist believed to show a young Gabriel Naninck, who would rise to be chief designer.

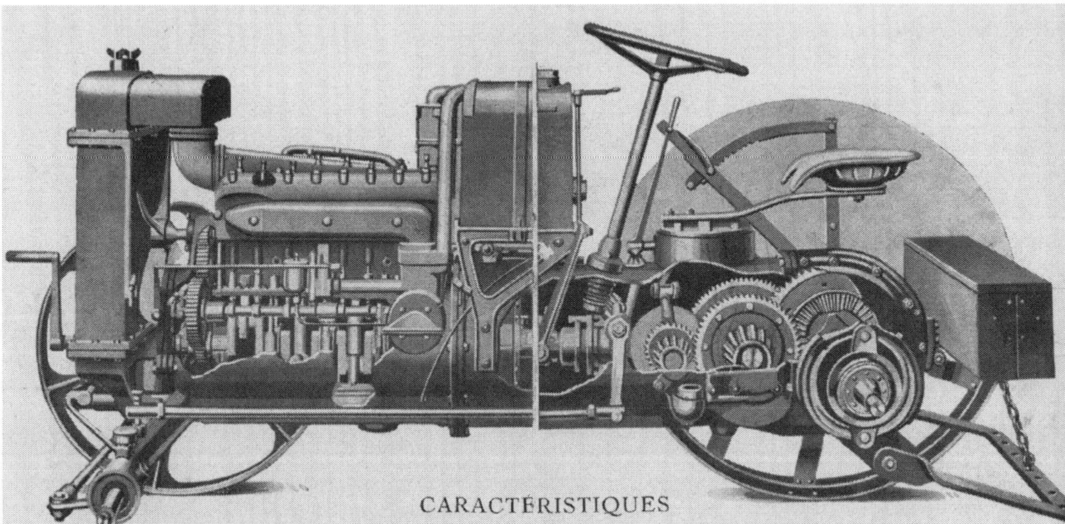

CARACTÉRISTIQUES

Châssis. — Constitué par un bâti carter étanche en trois parties formant T et dans lequel sont logés le moteur et tous les organes de transmission, qui sont à l'abri de l'eau, de la poussière et de la boue.
Suspension. — Le bâti repose sur l'essieu avant par l'intermédiaire d'une articulation à ressort formant ainsi une suspension élastique à trois points permettant au tracteur de prendre n'importe quelle position sans déformation ni fatigue.
Moteur. — Monobloc, quatre cylindres avec culasse amovible, fixé directement sur le bâti carter. Vilebrequin supporté par cinq paliers.
Graissage. — Sous pression par double pompe avec manomètre indicateur.
Refroidissement. — Par radiateur refroidi par ventilateur commandé par engrenages enfermés dans carter étanche. Circulation d'eau par pompe.
Carburateur. — Muni d'un épurateur d'air.
Régulateur. — A masses centrifuges sur l'admission des gaz.
Allumage. — Par magnéto à haute tension.
Embrayage. — Par cône léger en tôle d'acier entraîné par 6 segments amovibles garnis de Ferodo, montés à l'intérieur du volant (Brevet Austin).
Changements de vitesse et transmission. — Enfermés dans le bâti; arbres et engrenages en acier forgé, cémenté et trempé, noyés dans un bain d'huile et supportés par des paliers à rouleaux. Couronne arrière de grande dimension et à denture droite.
Freins. — Deux freins indépendants montés sur roues arrière.
Direction. — A vis et secteur enfermés dans le bâti.
Poulie. — Légère en acier, démontable en deux parties.
Carburants. — Essence ou pétrole sur demande.

□ 6 □

Empattement. —	1 m. 727
Voie — (axe en axe) Avant	1 m. 294
Arrière	1 m. 219
Largeur (extrême)	1 m. 550
Longueur	2 m. 769
Poids (en ordre de marche)	1.450 kg.
Rayon de virage	3 m. 35

Alésage des cylindres	95 m/m
Course des pistons	127 m/m
Puissance du moteur	29 HP
Régime du moteur	1.200 t.m.
Capacité des réservoirs à combus.	50 lit.
Roues. Avant	762 × 152 m/m
Arrière	1067 × 254 m/m

Se fait en 3 multiplications	Type Bourget	Petite vitesse avant	3 kilom. 200 à l'heure
		Grande vitesse avant	5 kilom. 500 à l'heure
		Marche arrière	2 kilom. 500 à l'heure
		Poulie (nombre de tours à la minute)	290
	Type St-Germain	Petite vitesse avant	4 kilomètres à l'heure
		Grande vitesse avant	7 kilomètres à l'heure
		Marche arrière	3 kilom. 250 à l'heure
		Poulie (nombre de tours à la minute)	360
	Type Buc recommandé pour les terres *très* dures.	Petite vitesse avant	2 kilom. 700 à l'heure
		Grande vitesse avant	4 kilom. 600 à l'heure
		Marche arrière	2 kilom. 100 à l'heure
		Poulie (nombre de tours à la minute)	245

□ 7 □

Cutaway view and specification of the early Liancourt R type with two gears but optional Bourget, St Germain and Buc ratios for different soil types.

L'AUSTIN LABOURANT ET HERSANT D'UN SEUL PASSAGE

La perfection du tracteur AUSTIN l'indique comme répondant tout aussi bien à la grande qu'à la petite culture parce qu'il s'adapte sans contredit à toutes les façons culturales tout en restant le plus économique.

La Bartherie, ce 29 Juillet 1923.

Monsieur,

Veuillez m'excuser si j'ai un peu tardé à vous donner mon appréciation sur le tracteur Austin ; je ne voulais le faire qu'en connaissance de cause.

Depuis un an que j'ai le tracteur, je n'ai eu qu'à me louer de mon acquisition.

Pour tous les travaux auxquels je l'ai employé, il ne m'a donné que de bons résultats. Sa parfaite stabilité m'a permis de labourer un champ accidenté à tel point que de l'avis de certains agriculteurs, propriétaires de tracteurs d'autres marques, c'était folie de vouloir le tenter. Le tracteur se jouant des pronostics a fonctionné avec une aisance parfaite.

Attelé à un extirpateur, il m'a donné les mêmes résultats qu'en labour, grâce à sa bonne adhérence.

L'an dernier, je l'ai essayé aux battages où il a émerveillé tout le monde.

Aux moissons, il ne s'est pas démenti ; il a remorqué une lieuse dans des penchants très prononcés avec une facilité parfaite et une consommation d'essence des plus réduites.

Recevez, Monsieur, toutes mes salutations.

Signé : J. HOLMIÈRE,
La Bartherie, par Venès (Tarn)

Ploughing and harrowing with three Austins in a single pass, quite possibly on Austin's own test fields, and using Bajac implements.

L'AUSTIN AVEC UN PULVÉRISEUR A DISQUES

Les pulvériseurs à grand travail de 32 à 40 disques sont remorqués par le tracteur AUSTIN.

besogne la plus utile, celle du « moment ». Et c'est bien celle-là dans notre métier que nous apprécions la première. A quoi bon herser d'avance si la pluie de la nuit nous fait recommencer le lendemain le même travail. Inutile de déranger des attelages, opération occasionnant toujours des pertes ; le tracteur, dès le matin, donnera l'avancée nécessaire.

Il herse un hectare à l'heure.

Il m'a traîné, sur un labour de défriche, deux Crosskills l'un derrière l'autre, faisant ainsi **7 hectares dans la journée pour deux passées de Crosskill.**

Il laboure facilement deux hectares par jour. Tout cela pour une **moyenne de consommation de 40 litres d'essence environ par jour.**

De plus, je reconnais à ce tracteur de **grandes facilités d'entretien, d'inspection, de manœuvre. Les organes sont souples, pratiques, bien protégés.**

Voilà mon appréciation, Monsieur, sur le tracteur que vous m'avez livré, ce ne sont que des éloges.

Recevez, je vous prie, l'assurance de mes sentiments distingués.

Signé : E. MORIN,
Agriculteur à Lieusaint (Seine-et-Marne)

Disc harrowing with a farmer late in 1920 who said that his Austin had been left running continuously for two months to avoid starting difficulties and freezing.

L'AUSTIN REMORQUANT UNE BATTEUSE

Le tracteur AUSTIN permet à l'entrepreneur de Battage de remorquer sa batteuse, sa presse à paille et ses divers accessoires dans tous les chemins.

Pechaudier, le 28 Juillet 1922.

Monsieur le Directeur
de la Société Anonyme Austin,

Je viens vous dire combien je suis satisfait de votre tracteur grande vitesse, avec lequel nous avons pu **labourer avec une charrue Olivier 12 pouces à deux socs dans une pente de 21 0/0, à 18 c/m de profondeur,** dans une terre **excessivement difficile** à travailler et **très dure.**

Le tracteur a **actionné** ensuite notre **batteuse Merlin à grand travail**, munie **d'un monte-gerbes** et **monte-paille**, avec la plus **parfaite aisance ;** devant ce résultat, je n'ai pas hésité à me rendre acquéreur d'une nouvelle batteuse.

Nous avons pu remorquer sur la route, **dans une pente de 6 0/0, un véritable train, composé de ma batteuse Merlin, de la locomobile et de deux charrettes chargées d'accessoires et de personnel, le tout pouvant être évalué à onze tonnes** et nous avons démarré dans la côte avec la plus parfaite aisance.

Je suis tout à fait étonné et émerveillé des résultats inespérés que nous avons obtenus et qui doivent se traduire pour votre marque par de nouvelles ventes dans la région.

Veuillez agréer, Monsieur le Directeur, l'assurance de mes sincères salutations.

Signé : LAPRADE,
Propriétaire à Pechaudier, par Cuq-Toulza (Tarn)

Hauling what appears to be a Garrett threshing tackle, for which Pilter was also French distributor.

L'AUSTIN ACTIONNANT UN CONCASSEUR

La faible consommation horaire du tracteur AUSTIN permet de l'utiliser à la ferme comme moteur fixe même pour actionner une machine ne demandant qu'une faible puissance : il reste, malgré tout « **économique** ».

Barquet, le 5 Août 1920.

Messieurs,

C'est avec plaisir que je viens vous adresser tous mes compliments au sujet du tracteur **AUSTIN** que vous m'avez vendu par l'intermédiaire de votre représentant Monsieur Bulle.

Je suis très content. J'espère et désire vivement que vous en vendiez d'autres dans la région.

Depuis que l'AUSTIN est entre mes mains, j'ai labouré, je coupe ma moisson et j'espère bientôt battre.

Tous mes remerciements.

Signé : ETIENNE Fils, Agriculteur à Barque
par Romilly-la-Puthenay (Eure).

An early R type with canopy shown running a mill from its belt pulley.

REMORQUAGE DE GRUMES

Le tracteur AUSTIN rend de très grands services dans les exploitations forestières.

Toulouse, le 7 Novembre 1921.

Monsieur le Directeur,

Comme je vous l'ai promis, je viens vous dire toute ma satisfaction du tracteur **AUSTIN**.

Seul, j'ai pu le mettre en marche, et dès le deuxième jour j'ai fait une moyenne de 2 hectares par jour travaillant à 3 charrues de 18 c/m. environ de profondeur ; quant à la dépense, naturellement, elle varie avec la nature du terrain.

En moyenne, j'ai dépensé **24 litres de pétrole** à l'hectare, mais dans certains terrains doux et ayant été bien travaillés auparavant, la dépense n'a été que de **20 litres à l'hectare**.

Toutes les personnes à qui j'ai causé et qui ont des tracteurs de différentes marques, ont été toutes étonnées de cette minime dépense.

Beaucoup sont venus voir travailler le tracteur et ceux qui ne le connaissaient pas ont été émerveillés de l'appareil.

En un mot, je ne regrette qu'une chose : c'est de ne pas l'avoir acheté plus tôt, et croyez que tout mes amis agriculteurs connaissent déjà mon opinion.

Croyez, Monsieur, à mes salutations les plus distinguées.

Signé : MONTALÈGRE
22, Rue St-Anne, Toulouse.

This photo appeared in an early French catalogue but shows an Austin plainly road registered in Britain.

L'AUSTIN REMORQUANT 14.700 KILOS

294 Sacs de 50 kilos — 52 kilomètres effectués chaque jour.

Travail fait quotidiennement par le tracteur AUSTIN depuis 4 ans, La grande puissance sous un faible volume du tracteur AUSTIN lui permet de remorquer de très grosses charges dans des conditions exceptionnellement économiques.

St-Sébastien. le 25 Décembre 1922.

Monsieur le Directeur,

Je dois vous annoncer avec fierté que je suis émerveillé du tracteur **AUSTIN** que je vous ai acheté au mois de Juillet dernier.

Nous avons **fait une campagne de battage merveilleuse, pas seulement une panne de 5 minutes** : tout le monde était surpris **de voir marcher cette batteuse si régulièrement et sans le moindre effort.**

Nous avons fait des labourages parfaits ; nous avons fait en terrain doux, un labourage à 0 m. 30 de profondeur, et une marche régulière.

Sur route, **nous avons fait des déménagements en traînant trois remorques, ou charettes chargées considérablement**. On est tout surpris de voir qu'un tracteur si petit puisse traîner un tel poids.

Je vous autorise à publier ma lettre, car elle est sincère.

Votre client tout à vous.

Signé : Eugène LAPRADE.
Saint-Sébastien (Tarn)

For four years this Austin hauled up to 14,700 kg for 52 km per day to a flour mill.

Above: Ploughing with a 'double brabant' made in Liancourt by Bajac.

Right: Sir Herbert Austin after his stint as an MP ended in 1924.

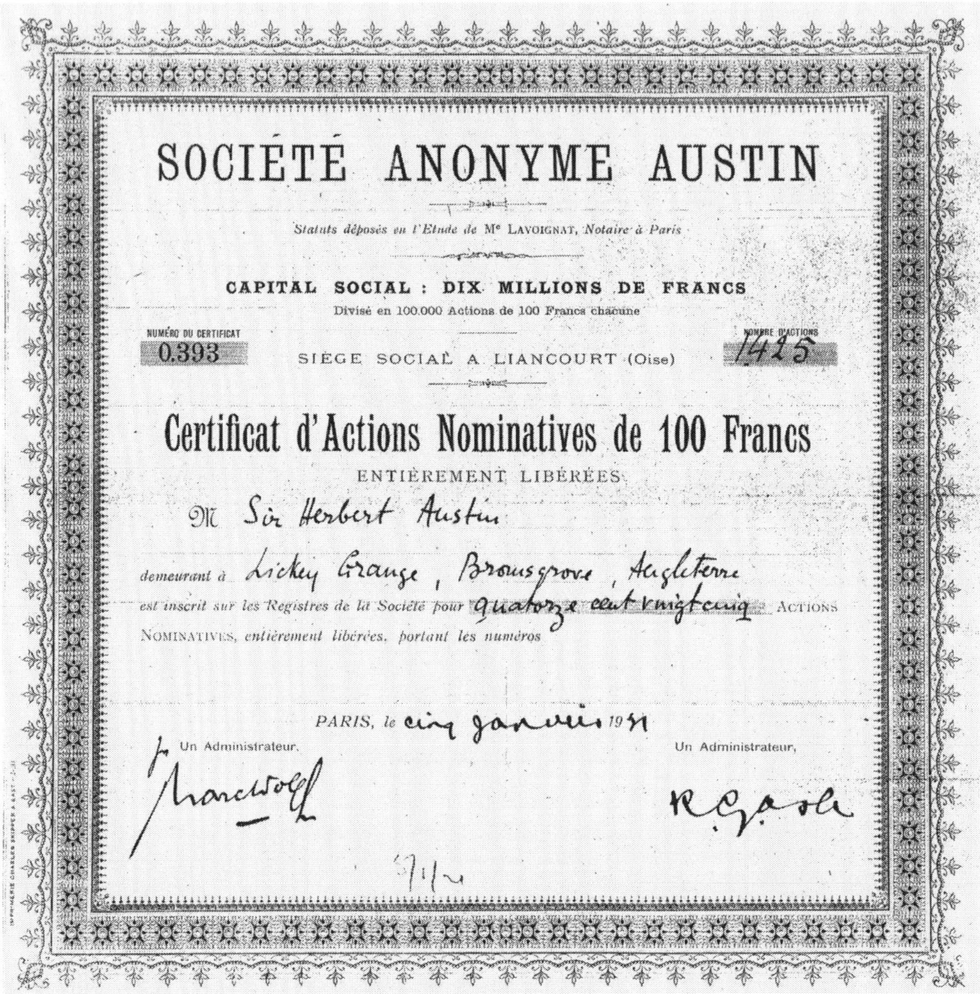

SOCIÉTÉ ANONYME AUSTIN

Statuts déposés en l'Etude de Mᵉ LAVOIGNAT, *Notaire à Paris*

CAPITAL SOCIAL : DIX MILLIONS DE FRANCS

Divisé en 100.000 Actions de 100 Francs chacune

NUMÉRO DU CERTIFICAT 0.393 — SIÈGE SOCIAL A LIANCOURT (Oise) — NOMBRE D'ACTIONS 1425

Certificat d'Actions Nominatives de 100 Francs

ENTIÈREMENT LIBÉRÉES

M Sir Herbert Austin

demeurant à Lickey Grange, Bromsgrove, Angleterre

est inscrit sur les Registres de la Société pour quatorze cent vingt cinq ACTIONS NOMINATIVES, *entièrement libérées, portant les numéros*

PARIS, *le* cinq janvier 1921

Un Administrateur, — Un Administrateur,

Some of the shares that Sir Herbert held in SA Austin, this batch dating from January 1921.

Demonstrating a disc harrow outside the Bajac works, on which can be read indistinctly the name of a wartime International Harvester tractor model.

This line block illustration showing a Bajac 'double brabant' is unfortunately undated, but shows the development with Austin of a power-driven chain implement lift.

Scrapyard view from the 1980s of a much-modified early Austin next to a 1950s Lanz. Note the Austin's square radiator core.

The French Austin radiator changed to this shape around 1923/4, though seldom had the Austin motif on the core.

AUS

22 **PAYEN**

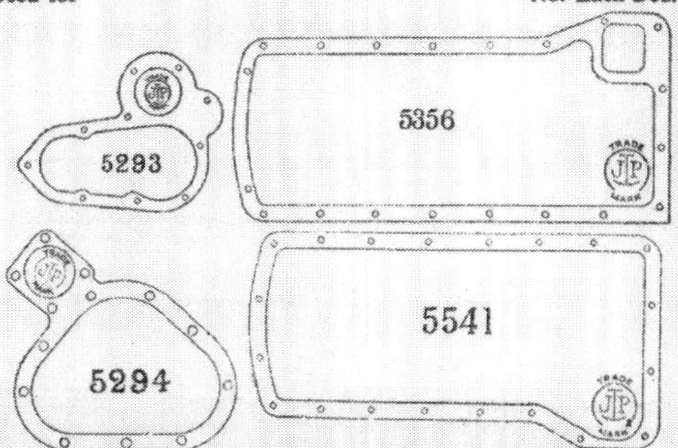

Used for			"J.P" No.	Price Each	per Doz.
Collection containing 1 set of Gaskets marked 1,			**DS2**	7/-	
" " " 2,			**OVS2**	12/-	
1, 2 Cylinder head			**472**	5/-	
Cylinder head, reinforced (recommended for old or rebored engines)			**1646**	6/-	
1, 2 Exhaust manifold, centre		2 off	**2018**	4d.	
1, 2 Exhaust manifold, front & rear	30×60·5	2 off	**30F2**	4d.	3/-
1, 2 Exhaust pipe	43×70	2 off	**43F2**	5d.	4/-
Exhaust pipe	Payenoid 43×70	2 off	**P43F2**	5d	4/-
Exhaust pipe "Blonot"	Payenoid 37×70		**P37CH2**	5d.	4/-
1, 2 Carburettor	Oakenstrong 31×53		**5185**	3d.	1/10
·2 Cylinder foot	Paper		**5017**	3d.	2/-
2 Water inlet	Oakenstrong		**5561**	3d.	2/-
Oil strainer cover	Oakenstrong		**5586**	6d.	4/6
2 Crankcase & camshaft, front end	Paper		**5285**	3d	1/8
2 Crankcase & camshaft, rear end	Oakenstrong		**5294**	1/-	11/-
2 Dynamo bracket	Paper		**5288**	2d.	1/-
2 Dynamo driving housing	Paper		**5289**	2d.	1/-
2 Magneto bracket	Paper	2 off	**5290**	2d.	1/4
2 Magneto to water pump	Oakenstrong		**5291**	4d.	3/-
2 Chain gear cover	Paper		**5292**	3d.	2/-
2 Top sump	Oakenstrong		**5356**	1/7	16/-
2 Bottom sump	Oakenstrong		**5541**	1/7	16/-
Gearbox front cover, 1921–33	Paper		**5284**	2d.	1/-
Gearbox front cover, 1934–38	Paper		**6904**	2d.	1/-
Gearbox rear cover, 1921–33	Paper		**5293**	3d.	2/-
Gearbox rear cover, 1934–38	Paper		**6903**	2d.	1/-
Gearbox top cover, 1921–33	Paper		**5287**	2d.	1/-
Gearbox top cover, 1934–38	Paper		**6817**	2d.	1/-
Worm gear, "Taxi"	Paper	2 off	**7896**	2d.	1/6
Worm drive housing front, "Taxi"	Paper		**7897**	2d.	1/-
Worm drive housing rear, "Taxi"	Paper		**7898**	2d.	1/-
Differential	Paper	2 off	**5286**	2d.	1/4
Rear axle	Felt 50×72	2 off	**5689**	9d.	7/6
Rear hub	Felt 47×92	2 off	**5639**	1/4	13/-
Collection containing 1 set of Gaskets marked 1,			**DS2**	7/-	
" " " " 2,			**OVS2**	12/-	

Austin 4 1920–29 20 H.P., Tractor, P4, P5, P6, 16 cwt.

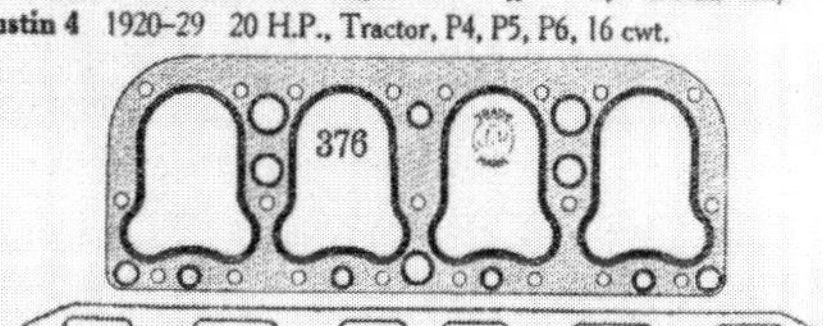

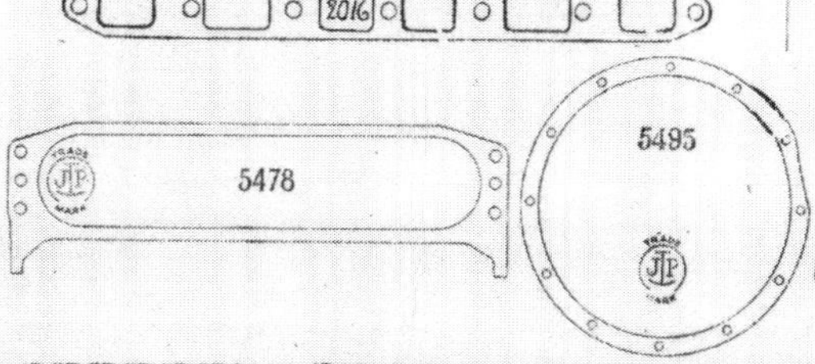

Used for			"J.P" No.	Price Each	per Doz.
Collection containing 1 set of Gaskets marked *,			**S35**	12/6	
*Cylinder head			**376**	8/8	
Cylinder head, reinforced (gasket bore 100 m/m.) (recommended for old engines)			**1821**	8/8	
*Exhaust manifold			**2016**	2/4	
*Exhaust pipe	57·5×61	3 off	**57SQ2**	7d.	
Exhaust pipe	Payenoid 57·5×61	3 off	**P57SQ2**	7d.	
*Induction pipe	40×58	2 off	**40F1**	5d.	4/-
*Carburettor	Oakenstrong 36·5×65·5		**5186**	3d.	2/-
*Cylinder foot	Paper		**5478**	3d.	2/-
Differential	Paper	2 off	**5495**	4d.	3/2
Rear hub	Felt 54×105	2 off	**6189**	1/4	13/-
Collection containing 1 set of Gaskets marked *.			**S35**	12/6	

Austin 4 Tractor (French) 1925–29 25 H.P. Model BD28

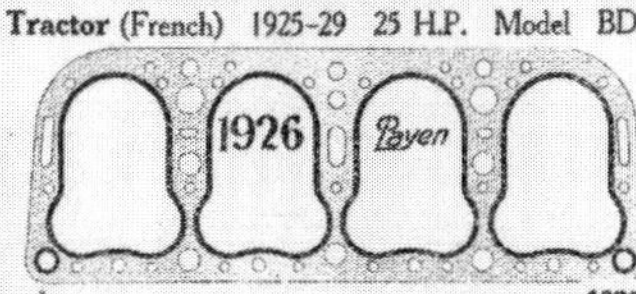

Cylinder head		**1926**	9/-

Austin 6 1931–36 (13·9 R.A.C.), 10 CV., "Light Twelve Six," "Sports 15·9 H.P.," 10 cwt.

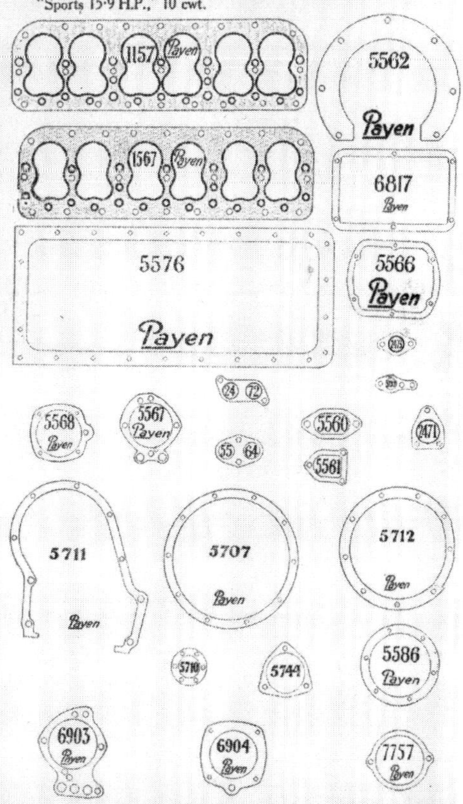

Collection containing 1 set of Gaskets marked 1,		**DS33A**	**8/10**
" " " " 2,		**OVS33A**	**11/-**
" " " " 3,		**DS149**	**8/10**
" " " " 4,		**OVS149**	**11/6**

Please obtain your supplies from Payen Stockists

Interesting contrast in cylinder head gaskets between British and French tractors from the Payen guide. The BD type was actually from later in the 1920s.

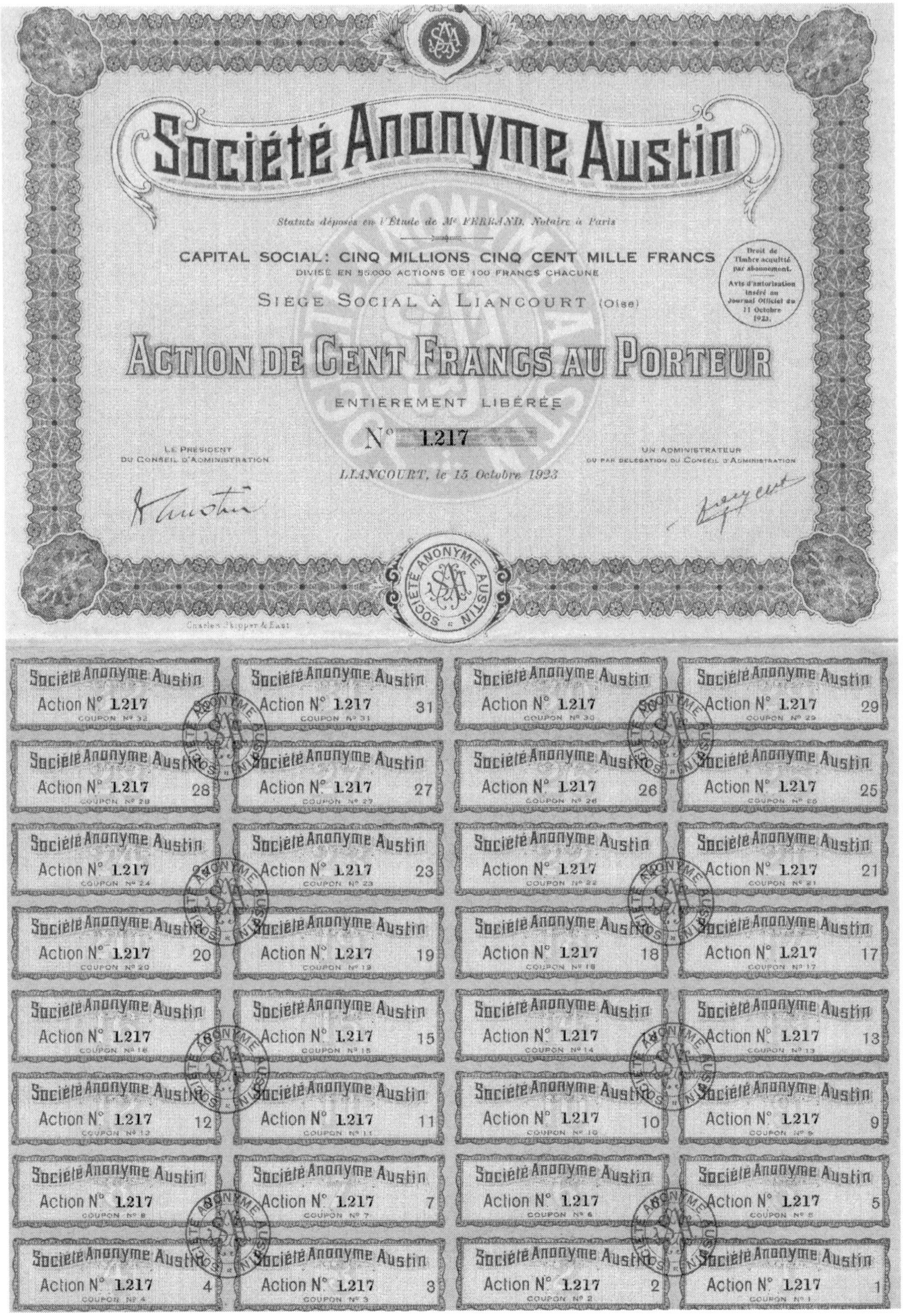

Stylish 1923 SA Austin share certificate. It was part of an attempted 5.5 million franc flotation.

SOCIETE ANONYME AUSTIN.

BALANCE SHEET.

Complimentary Copy.

AS AT 31st DECEMBER, 1927.

ASSETS.

	£ s. d.	Francs.	Francs
Cash on Hand and at Banks :—			
In francs	319 5 1	39.593,79	
In Foreign Currencies (Sterling, Dollars and Swiss Francs)	807 9 3	100.141,53	
	1,126 14 4		139.735,32
Bills Receivable	26 11 8		3.296,65
Sundry Debtors :—			
Customers	5.057 2 5	627.184,24	
Suppliers' Debit Balances	82 15 0	10.262,57	
Payments and Rent in advance	929 3 4	115.235,40	
	6,069 0 9		752.682,21
Stock of Merchandise :-			
Raw Materials and Stores	19,894 14 5	2.467.343,37	
Work in Progress	6,511 6 3	807.532,71	
Tractors	4,733 13 9	587.071,98	
Tools	1,554 18 7	192.842,41	
	32,694 13 0		4.054.790,47
Fixed Assets (*Less* Depreciation):—			
Land and Buildings	12,493 3 10	1.549.405,70	
Plant Installation and Motor Lorries	17.706 1 11	2.195.910.14	
Jigs and Templates	3,170 10 2	393.206,41	
Furniture	476 7 5	59.079,45	
	33,846 3 4		4.197.601,70
	£73,763 3 1		Frs. 9.148.106,35

LIABILITIES.

	£ s. d.	Francs.	Francs.
Share Capital	44,347 13 9		5.500.000,00
Legal Reserve	344 12 4		42.739,31
Extraordinary Reserve	4,434 15 4		550.000,00
Sundry Creditors :—			
Suppliers	5,312 3 7	658.816,52	
Customers' Credit Balances	1,239 12 8	153.739,28	
Outstanding Expenses	1,620 3 2	200.932,40	
Sir Herbert Austin	2,800 0 0	347.256,00	
The Austin Motor Co. Ltd. Franc Account	2,736 5 11	339.355,20	
	13,708 5 4		1.700.099,40
Mortgage Debenture	7,606 3 7		943.318,20
Provision for Doubtful Debts	260 13 10		32.331,15
Profit and Loss Account :—			
Net Profit brought forward	2,112 18 10	262.046,96	
Add— Net Profit for the year 1927	948 0 1	117.571,33	
	3,060 18 11		379.618,29
	£73,763 3 1		Frs. 9.148.106,35

NOTE :—The figures have been converted into Sterling at the Rate of Exchange at 31st December, 1927, namely 124·02 Francs to the £

The only SA Austin balance sheet to have come to light so far shows a relatively healthy situation.

LE TRACTEUR
Austin
A HUILE LOURDE
RÉALISE
60%
D'ÉCONOMIE
Sur la dépense en carburants
Le tracteur AUSTIN a résolu le problème que se pose tout agriculteur soucieux de l'avenir de son industrie. C'est celui qui répond le mieux aux besoins de l'Agriculteur Français.
A tous les Travaux de la ferme
LABOURS
DEFONÇAGES
MOISSONS
BATTAGES
REMORQUAGES
Maximum d'Adhérence au sol
C'est celui qui vous fera toujours le travail du moment.
C'est celui qui diminuera vos dépenses et augmentera le rendement de vos terres.
Il vous assure ses qualités de
SIMPLICITÉ
PUISSANCE
ROBUSTESSE
DURÉE
ECONOMIE
Fabrication Française
Références dans toutes Régions - Demandez le Catalogue illustré gratuit série AM à
SOCIÉTÉ ANONYME AUSTIN
5, Rue Cardinal Mercier, PARIS (9) — Usines à LIANCOURT (Oise)
AUSTINTRAC-PARIS

Undated publicity for heavy oil (paraffin) powered Austin, noting French manufacture.

1928 advertisement states that the Austin tractor is the most common or widespread in France and comes from the only factory making tractors exclusively.

Early 1929 advertisement showing the usual 15/25, by then called the B028.

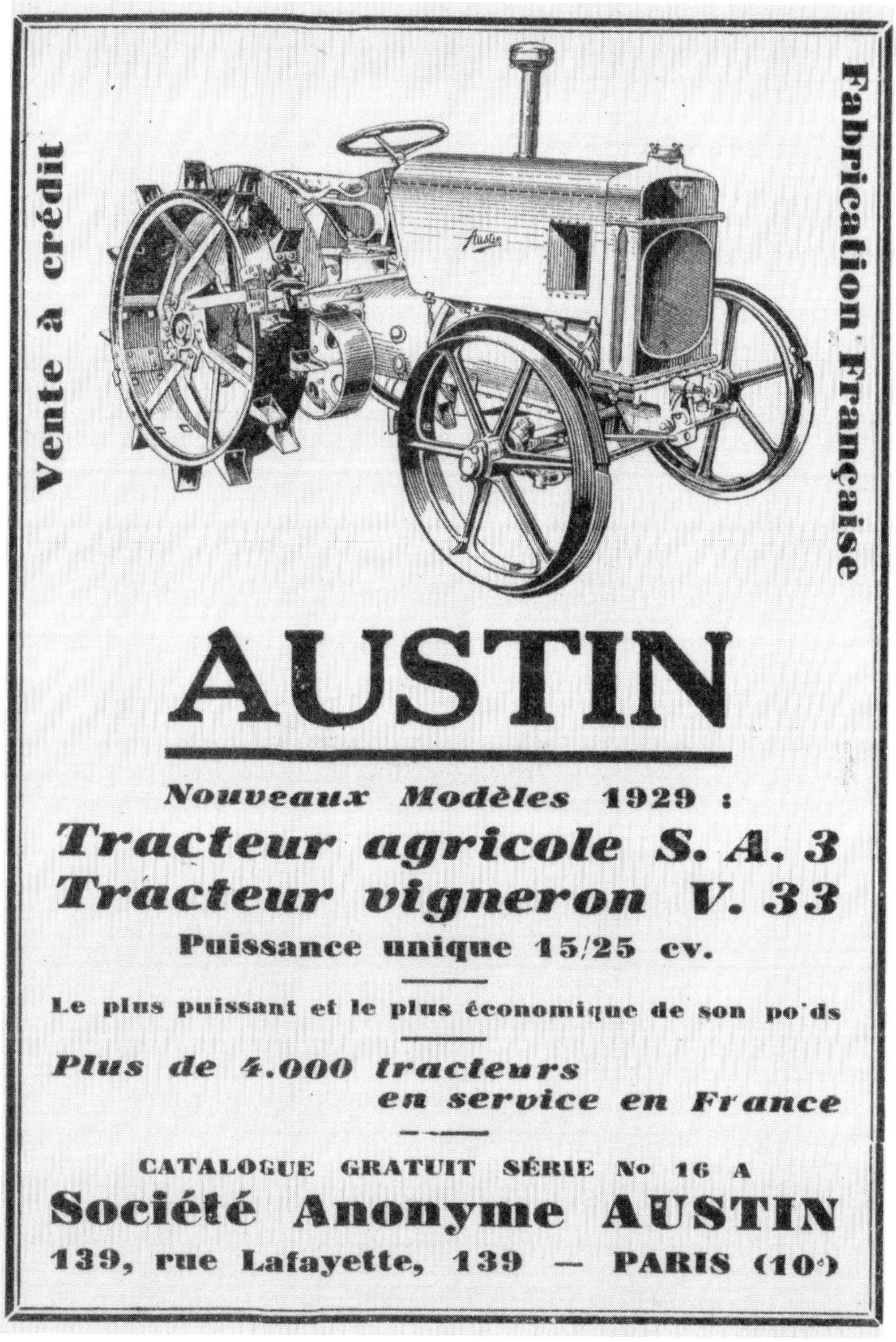

By March 1929 the 15/25 was called the SA3 or V33 vineyard model, which were claimed to be the most powerful and economical for their weight class.

Starting an SA3 on the upstroke with thumb on same side as fingers to avoid injury in the case of backfiring that could happen in picture to left.

Splitting an SA3 for major overhaul was not as easy as it looks because the halves had to be wheeled apart.

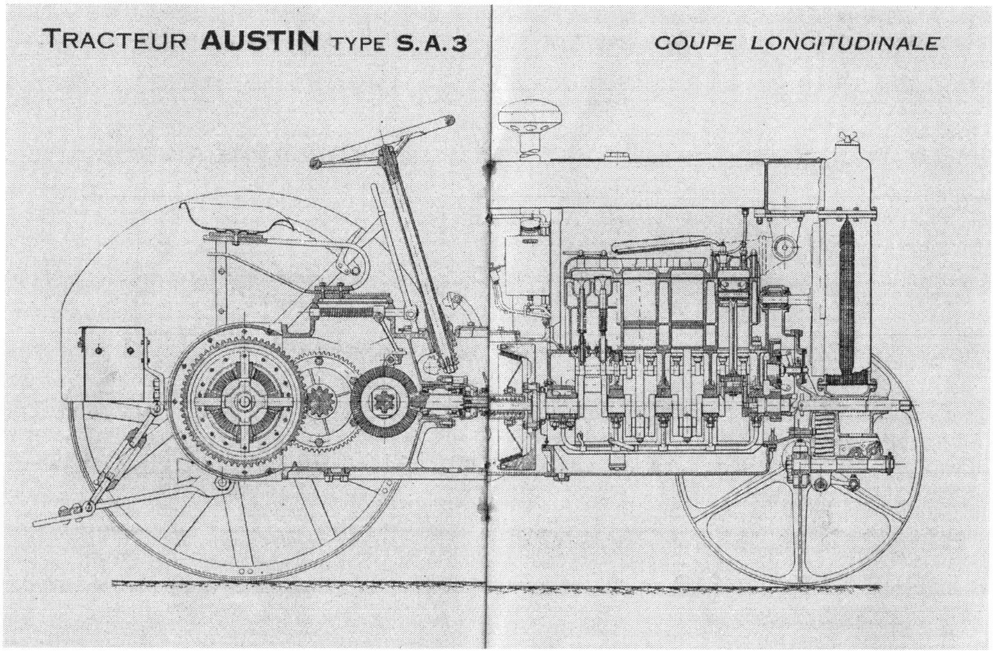

Cross-section through the SA3 shows it to have been an improved R type.

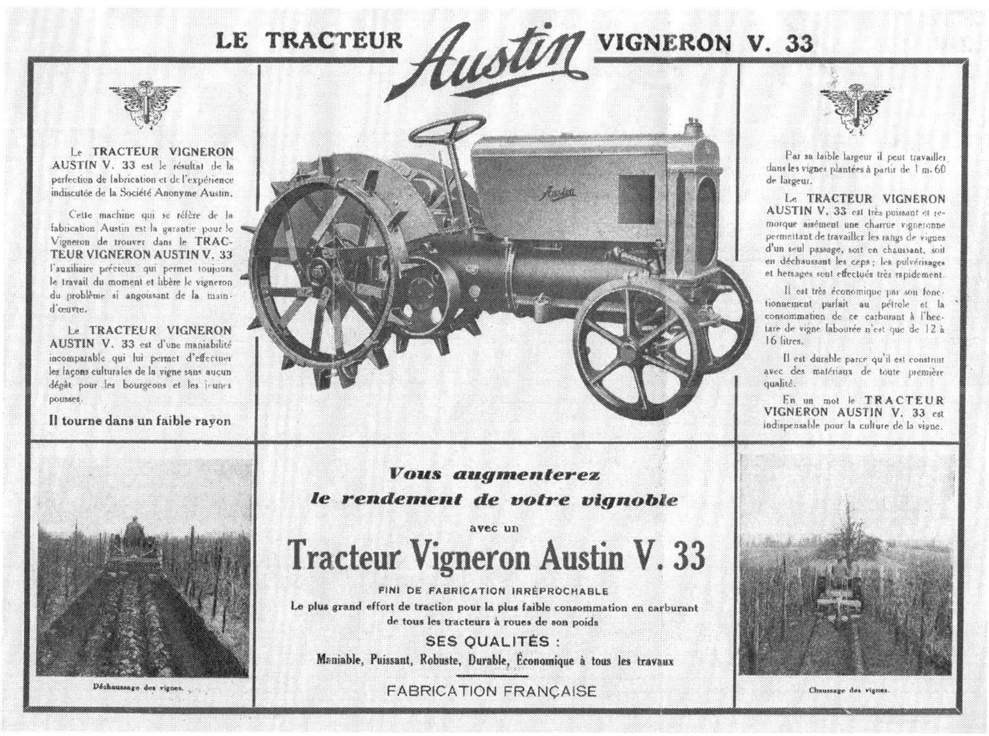

January 1929 leaflet for V33 which was only 95 cm wide to fit between vine rows. Note the detachable lugged field bands.

The author's V33 with bonnet removed for adjustments sets out on its first field trials in forty years.

Minus its field bands, the author's V33 serial 3611–10 attends Yesterday's Farming event in Somerset.

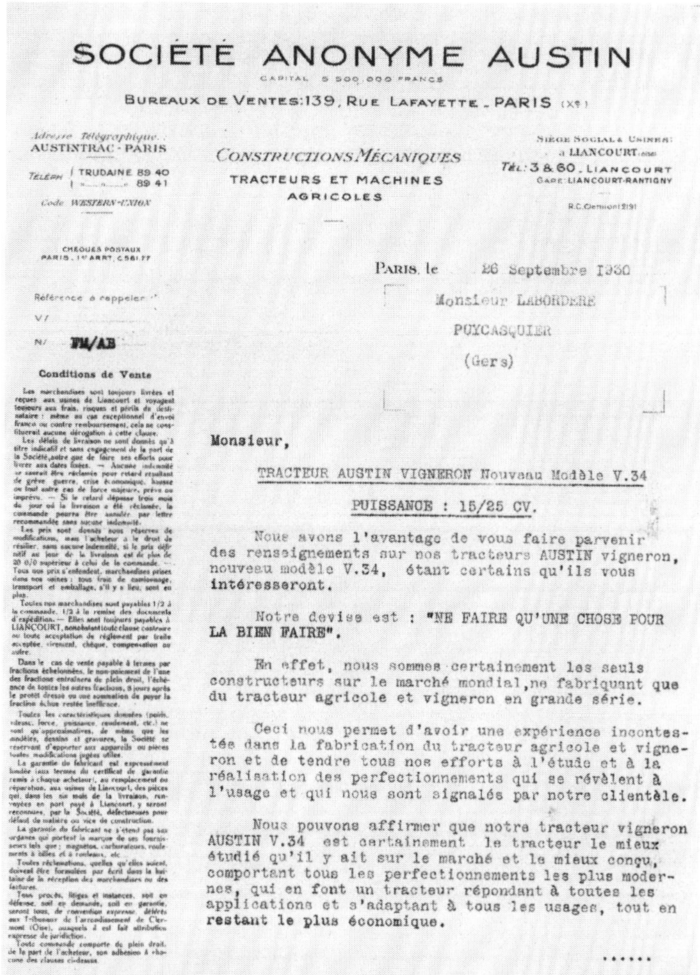

SOCIÉTÉ ANONYME AUSTIN

CAPITAL 5 500 000 FRANCS

BUREAUX DE VENTES: 139, RUE LAFAYETTE _ PARIS (X^e)

Adresse Télégraphique : AUSTINTRAC - PARIS

TÉLÉPH. TRUDAINE 89 40 / 89 41

Code WESTERN-UNION

CHÈQUES POSTAUX PARIS. 1er ARRT. C561.77

CONSTRUCTIONS MÉCANIQUES

TRACTEURS ET MACHINES AGRICOLES

SIÈGE SOCIAL & USINES : à LIANCOURT (Oise)

TÉL : 3 & 60. LIANCOURT

GARE : LIANCOURT-RANTIGNY

R.C. Clermont 2191

Référence à rappeler :

V/

N/ FM/AB

PARIS, le 26 Septembre 1930

Monsieur LABORDERE

PUYCASQUIER

(Gers)

Conditions de Vente

Les marchandises sont toujours livrées et reçues aux usines de Liancourt et voyagent toujours aux frais, risques et périls du destinataire : même au cas exceptionnel d'envoi franco ou contre remboursement, cela ne constituerait aucune dérogation à cette clause.

Les délais de livraison ne sont donnés qu'à titre indicatif et sans engagement de la part de la Société, autre que de faire ses efforts pour livrer aux dates fixées. — Aucune indemnité ne saurait être réclamée pour retard résultant de grève, guerre, crise économique, hausse ou tout autre cas de force majeure, prévu ou imprévu. — Si le retard dépasse trois mois du jour où la livraison a été réclamée, la commande pourra être annulée par lettre recommandée sans aucune indemnité.

Les prix sont donnés sous réserves de modifications, mais l'acheteur a le droit de résilier, sans aucune indemnité, si le prix définitif au jour de la livraison est de plus de 20 0/0 supérieur à celui de la commande. — Tous nos prix s'entendent, marchandises prises dans nos usines : tous frais de camionnage, transport et emballage, s'il y a lieu, sont en plus.

Toutes nos marchandises sont payables 1/2 à la commande, 1/2 à la remise des documents d'expédition. — Elles sont toujours payables à LIANCOURT, nonobstant toute clause contraire ou toute acceptation de règlement par traite acceptée, virement, chèque, compensation ou autre.

Dans le cas de vente payable à termes par fractions échelonnées, le non-paiement de l'une des fractions entraînera de plein droit, l'échéance de toutes les autres fractions, 8 jours après le protêt dressé ou une sommation de payer la fraction échue restée inefficace.

Toutes les caractéristiques données (poids, vitesse, force, puissance, rendement, etc.) ne sont qu'approximatives, de même que les modèles, dessins et gravures, la Société se réservant d'apporter aux appareils ou pièces toutes modifications jugées utiles.

La garantie du fabricant est expressément limitée (aux termes du certificat de garantie remis à chaque acheteur), au remplacement ou réparation, aux usines de Liancourt, des pièces qui, dans les six mois de la livraison, renvoyées en port payé à Liancourt, y seront reconnues, par la Société, défectueuses pour défaut de matière ou vice de construction.

La garantie du fabricant ne s'étend pas aux organes qui portent la marque de ses fournisseurs tels que : magnétos, carburateurs, roulements à billes et à rouleaux, etc.

Toutes réclamations, quelles qu'elles soient, doivent être formulées par écrit dans la huitaine de la réception des marchandises ou des factures.

Tous procès, litiges et instances, soit en défense, soit en demande, soit en garantie, seront tous, de convention expresse, déférés aux Tribunaux de l'arrondissement de Clermont (Oise), auxquels il est fait attribution expresse de juridiction.

Toute commande comporte de plein droit, de la part de l'acheteur, son adhésion à chacune des clauses ci-dessus.

Monsieur,

TRACTEUR AUSTIN VIGNERON Nouveau Modèle V.34

PUISSANCE : 15/25 CV.

Nous avons l'avantage de vous faire parvenir des renseignements sur nos tracteurs AUSTIN vigneron, nouveau modèle V.34, étant certains qu'ils vous intéresseront.

Notre devise est : "NE FAIRE QU'UNE CHOSE POUR LA BIEN FAIRE".

En effet, nous sommes certainement les seuls constructeurs sur le marché mondial, ne fabriquant que du tracteur agricole et vigneron en grande série.

Ceci nous permet d'avoir une expérience incontestée dans la fabrication du tracteur agricole et vigneron et de tendre tous nos efforts à l'étude et à la réalisation des perfectionnements qui se révèlent à l'usage et qui nous sont signalés par notre clientèle.

Nous pouvons affirmer que notre tracteur vigneron AUSTIN V.34 est certainement le tracteur le mieux étudié qu'il y ait sur le marché et le mieux conçu, comportant tous les perfectionnements les plus modernes, qui en font un tracteur répondant à toutes les applications et s'adaptant à tous les usages, tout en restant le plus économique.

......

The V33 had become the 1.1-metre-wide V34 by the time that SA Austin director A. Kauffmann sent a five-page sales letter to a prospect on 26 October 1930.

Austin produced a blue and white enamel sign to publicise its tractors. One is shown here in a French agricultural museum behind International Titan and cross-motor Case tractors.

Assortment of Austins seen in collections and farmyards. The preserved 1930 example in blue with red wheels is on display with Berliet lorry at 46250 Cazals/Montclera. The one on rubber tyres is all red. The two Austins together were seen near Hay-on-Wye.

FABRIQUÉ EN FRANCE
Austin
TRACTEURS INDUSTRIELS
AUSTIN
IPE 29 TRACTEUR
MONTÉ SUR PNEUMATIQUES
PUISSANCE 15 x 25 CV.
PUISSANCE IMPOSABLE 14 CV.

There was an Austin industrial tractor in 1925 with three forward gears and rubber tyres. This is the 1930 IBA28 version.

Les Etablissements REBOUL & GAY
sont Agents Exclusifs pour les Bouches-du-Rhône et le Var
des TRACTEURS Cletrac
Le Tracteur à chenilles qui a fait ses preuves
Primé
par le Ministre de la Guerre
Primé
par le Ministre de la Guerre
Une prime de 6.100 francs payable en trois annuités est accordée par le gouvernement à chaque propriétaire de tracteur "CLETRAC" K ou K E, muni d'un équipement spécial

Austin's agent in Marseilles was Reboul & Gay, who also handled tractors by Lanz, Centaur and Cletrac. The latter American make had been promoted in Britain by SA Austin director H. G. Burford. In 1930 it cost 43,500 francs while the 15/30 Lanz cost 46,300 francs against the home grown 15/25 Austin at 28,500 francs.

Art deco catalogue promoting Bergougnan solid rubber-tyred roadbands for numerous applications, including this Austin.

Finger mowers by Goguet & Mermet of Romans enabled Austins to mow one hectare per hour.

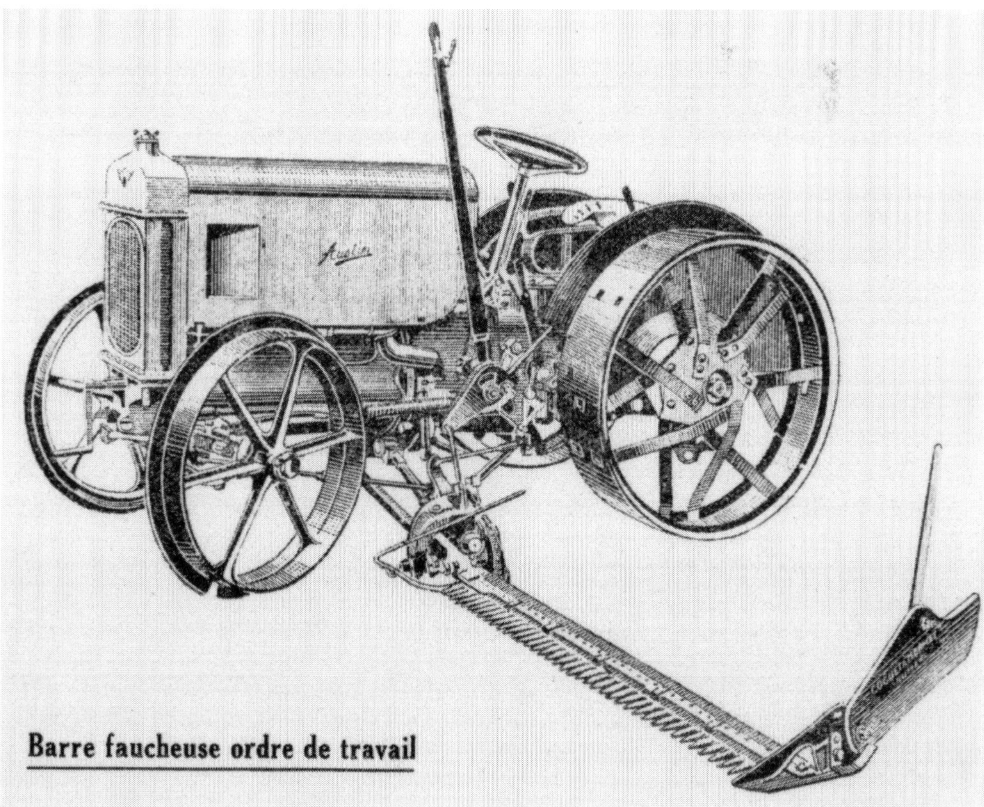

Although marked Austin, this 1930 picture shows an untypical radiator and bulkhead. Reboul & Gay had delivered over 700 Austins to farmers in Provence.

Listed for 1930 were the mechanically similar 15/25 DE30 and V34, and the 22/35 hp, which had a similar motor but with 105 mm instead of 95 mm bore.

Charrues et Instruments pour la Vigne

Cultivateur à dents flexibles REBOUL & GAY

pour tracteurs AUSTIN et CLECTRAC

Le cultivateur permet de travailler dans les vignes dont les rangées sont écartées de 2 mètres et plus. Il est très rapidement réglable en largeur par l'adjonction ou la suppression de barres transversales qui portent des dents seulement à leurs extrémités. Solide et rustique, il est d'une mise au point parfaite pour un travail intensif des façons culturales. Dans les vignes plantées de 2 m. à 2 m. 50 on peut cultiver facilement l'hectare à l'heure avec un tracteur AUSTIN ou CLETRAC pour lesquels nous l'avons étudié spécialement.

Sur les ressorts de ce cultivateur on peut monter des socs dits rasettes ou ailes d'hirondelles.

Cette gravure représente l'appareil relevé par le relevage vendu séparément et qui se pose sur le tracteur pour relever tous instruments attelés au tracteur.

Prix du Cultivateur comprenant 15 dents : 1.200 Francs (avec ses ferrures de montage)

Nous adaptons notre cultivateur à n'importe quel système de relevage

Relevage Automatique REBOUL & GAY

pour tracteur AUSTIN Vigneron

Ce relevage a été construit pour permettre de relever les instruments au bord des raies de vignes afin de manœuvrer aisément avec le tracteur, car pour faciliter les manœuvres, quand les fourrières sont très courtes ou en pointes, il faut que l'instrument dont on se sert soit relevé pour permettre de faciliter la marche arrière.

Ce relevage automatique permet également de pouvoir se servir d'instruments plus courts que les différents appareils munis d'un relevage monté sur ces dits appareils.

Ne peut se monter que sur tracteur Vigneron **AUSTIN**.

Prix de ce relevage : 2.800 francs

Ce relevage reste le plus simple, ne peut pas se détériorer. Dure de nombreuses années sans aucune réparation.

Nous le recommandons particulièrement aux propriétaires qui ont supprimés dans leurs vignes, les labours et ne font que des façons culturales, en employant notre cultivateur à dents flexibles et pour lequel il convient merveilleusement.

Fonctionne sur de nombreux tracteurs AUSTIN depuis plusieurs années à la satisfaction générale de ceux qui l'emploie.

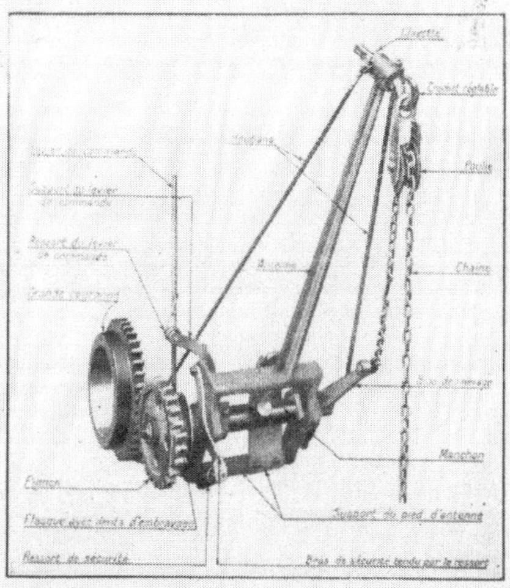

Reboul & Gay mechanical implement lift for Austins and Cletracs.

The V90 versions of the V34 was only 90 cm wide and did without a belt pulley. Like its contemporaries it had a new high-mounted and visible air cleaner.

Above and below: Gone was the traditional Austin script on the 1931 models, replaced by a bolder painted name. This is the 17-mph industrial model on giant pneumatics called an IPE29 (the solid-tyred version was IBA28).

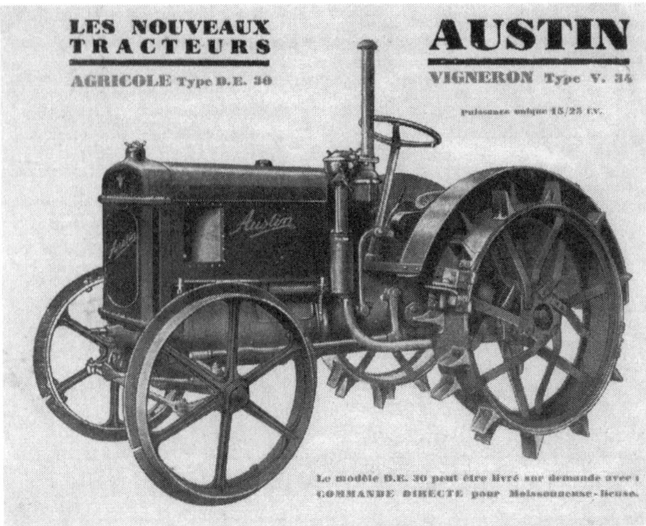

LES NOUVEAUX
TRACTEURS
AGRICOLE Type D.E. 30
AUSTIN
VIGNERON Type V. 34
Puissance unique 15/25 CV.
Austin
Le modèle D.E. 30 peut être livré sur demande avec :
COMMANDE DIRECTE pour Moissonneuse-lieuse.

A DE30 from just before the change of side script. Note air cleaner, detachable rims and message that it could be driven direct from reaper binder with extended controls.

Austin
15/25
Standard FARM TRACTOR
1930 MODEL
Equipped for Paraffin or Petrol.
Supplied complete with Declutchable Pulley and Tools.
INCLUSIVE PRICE.
Power Take-off supplied as Accessory.
On Show—
ROYAL SHOW – Manchester Meeting
July 8th - 12th.
STAND 63 - Machinery in Motion Section
Enquiries :
"AUSTIN TRACTORS," Messrs. R. & J. PARK, LTD.,
Dominion House, Thames Road, Chiswick, LONDON, W.4

The DE30 advertised in Britain in July 1930, where it was known simply as a 15/25. It had double air filtration and a Tecalemit oil filter.

Contemporary Austins were generally painted pale blue but a few exist in yellow with red wheels.

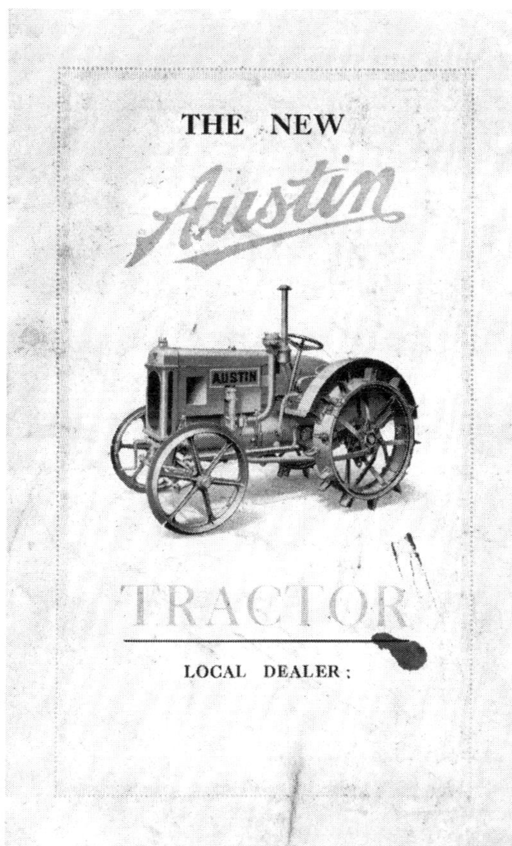

This battered catalogue for the 25/35 was dragged from a bonfire prior to Parkhouse blacksmith's 1990s clearance sale in Horton, Somerset. Confusingly the tractor is described as a 15/28, though the catalogue describes that 30/34.8 hp was recorded on an official Oxford test in May 1931.

This sectioned Austin was at the Paris agricultural show of January 1930. The other picture from the catalogue shows a tractor by Bauche – the make represented by Maison Pilter after its falling-out with Liancourt. It also represented Case from the later 1920s.

Chapter 5

Diesel and Beyond

Reliable and economical diesels swept to prominence in 1930 among commercial vehicle makers and Austin tried versions of its engines starting on petrol that could be switched to fuel injection and higher compression when warm enough to run on diesel fuel. By then the engines had overhead valves and, in late 1932, an ante-chamber design that gave them full diesel status and, according to Austin, no need for pre-heating. They were probably referring to semi-diesel hot bulb rivals that required an outside blow lamp to aid starting. In fact, the Austins did have provision for wick or fuse-type burning tapers to be inserted into their induction manifolds.

A combination of decompressors and geared starting handle combined with heaters made starting feasible. However, on cold days owners cursed the way burners went out before all could be lit. Austin owners today tend to use phosphorous-type Hatz burners or electric heater plugs.

There was mention of a new 7.2 litre four-cylinder kerosene model in 1933, which implied petrol-paraffin but maybe was meant to infer diesel and, indeed, a diesel 45/55 arrived in 1934. Despite Sir Herbert's apparent enthusiasm for the French productions, there were none at the 1934 nor 1936 Royal Agricultural Show.

Sir Herbert made a big donation to cancer research in 1934, having been on the board of management of Birmingham Hospitals since 1918. Then in 1936 he donated a quarter of a million pounds to Lord Rutherford's radiology research at Cambridge and, because of this and his other generosity, was ennobled as Lord Austin of Longbridge KBE. His factory at Longbridge was registering record sales and exports, and employed 18,500 at the time.

Austin started an 11.5-acre Shadow aircraft factory at nearby Crofton Hacket and decided to return to lorries after an absence of over twelve years. Both Foden and Arran had built Austin 20 petrol-engined lorries in the 1930s and, with the success of dieselised versions in France, there was talk of trying Austin's own diesel lorries in Canada in late 1936. There must have been problems with upping the revs from the tractor's governed 1,200 rpm as, when Austin lorries finally appeared in late 1938, they were six-cylinder petrol-powered.

Since 1938 Lord Austin, aged seventy-two, had been stepping back from some of his many commitments. He appointed Leonard Lord from rivals Morris as Managing Director of the Austin Motor Co. in 1938. Lord had been with Vickers, Daimler, and was with the Hotchkiss engines factory in Coventry when William Morris bought it in 1923.

Lord Austin had been selling his shares in SA Austin to Yugoslavian citizen Robert Rothschild (born in Zagreb in 1895), who handled French sales of German Hanomag tractors with his Yugoslavian brother-in-law Milos Celap. By April 1939, when only the 22/35 and 45/55 tractors were listed (with slightly enlarged capacity of 4.45 and 7.28 litres), Rothschild held ninety-one per cent of SA Austin's share capital, which stood at 3 million francs, down from 5.5 million in the early 1930s. A February 1938 press release had referred to 45/55 tractors being exported to England and the British dominions.

In June 1940, German invaders took over the factory (where employment had recently stood at 180 men), and Rothschild handed over most of his shares to Celap. Tractor production continued into 1942 but German authorities considered that the works still belonged to Rothschild, a Jew (though Celap was not), and seized the premises. It was used by the Krupp steel and armaments company (which had introduced tractors in the 1930s as a likely cover for military vehicles). Krupp, who had also stolen the Elmag factory in Alsace, used both to make war material, including grenades and Daimler-Benz DB-10 Blockwagens at Liancourt in the most terrible conditions using slave labour. The details emerged from the post-war Nuremberg Military Tribunal, at which Alfred Krupp was sentenced to twelve years jail and forfeiture of all property.

Robert Rothschild and his wife went to live near the Vincent family, who were the Austin tractor agents near Lyon, but he was eventually betrayed to the pro-German Vichy government and sent to a camp for Jews at Drancy in north-east Paris. From there he was deported to Auschwitz where, soon after 7 March 1944, he was murdered. More revelations about the cruel and dreadful events at Liancourt and other Krupp plants are contained in dozens of pages of trial depositions, which can be referred to on the internet. Included above is only a very brief account.

Fortunately, Lord Austin knew none of it, though he saw the effects of the war when Longbridge was bombed. He attended the funeral of six victims, contracted a chill that developed into pneumonia, and died on 23 May 1941.

In 1945 the millionth Austin vehicle was made and the Longbridge factory had grown from its original 2.5 acres to 100. At Liancourt, Celap and his sister returned to pick up the pieces and would live there into the 1960s. Around forty men were employed mending tractors that had come in during the war years for repair. There must have been around 10,000 in use by then (the author has seen a mid-1930s diesel with serial number above 9,000 and a 7.2 litre type numbered 10010), so demand for spares was presumably considerable. Some limited new production took place and there were said to be two new types in preparation. However, they did not prosper, and Austin vanished from sales lists and agricultural shows, though catalogues were still being issued around 1950 when 25 and 35 hp petrol and diesel models were listed, with the 7.2 litre diesel type by then called a 50D. The official French tractor parc figures of June 1950 recorded 25,323 Renaults in top position followed by SFV with 6,967, MAP with 3,535 and Latil with 1,631. Among imports, International Harvester accounted for 22,428, followed by the new Ferguson at 14,487, Allis-Chalmers at 4,720, Case with 3,336, Lanz with 2,753 and Ford with 2,646. Austin should have been well up in one or the other group, but instead vanished mysteriously in the 20,000 'diverse and indeterminate', which is roughly the number that one source ambitiously gave for total Austin tractor production – probably based on the 2,000 a year potential figure quoted at the outset.

The works were kept busy making engines; not the local Tosellos, but the better-known Bernards, though whether this happened before or after the 1953 closure referred to in the reproduced letter is unclear. One source claimed 1951 as the date for Bernard at Liancourt. 1957 marked the end of Liancourt's traditional shoe manufacture, production having fallen from 10,000 pairs per day in 1914 to 300 per day.

Some of Austin's subsequent involvements with agriculture are illustrated in this chapter, including the Bristol crawler that used Austin engines from 1942 and was adopted by Austin distributor HA Saunders in 1961. The Liancourt factory had become home to Floquet Monopole piston rings and engine parts in the mid-1950s. The neighbouring Bajac plough works closed in 1962 and became Siccardi, making components for the Massey-Ferguson factory that had opened at nearby Beauvais in 1957. The Albaret works was bought by Caterpillar in 1985 and closed down in 2015. Siccardi and Floquet Monopole had both departed by then and their factories been flattened.

As a postscript comes the puzzling enigma of Harry Ellard's attempts to reactivate Austin tractors in Britain. In 1984 his collection of classic cars – many of them Meadows-engined Lagondas – was sold by auction and among them were many tons of Austin tractor parts. Ellard was an engineer with a big shareholding in the Meadows engine business. He also made casting patterns and cores for Austins and found time to farm in the Cotswolds where he had used Austins and Cletracs. He is said to have intended to make British Austin tractors in the 1930s, which begs the question – why did Austin sell the spares to him and not Liancourt? Perhaps the vexed question of slow payment had not been forgotten.

Late in 1932 Austin produced its first diesel models – the 16/28 and 22/36, based on their two sizes of petrol tractor.

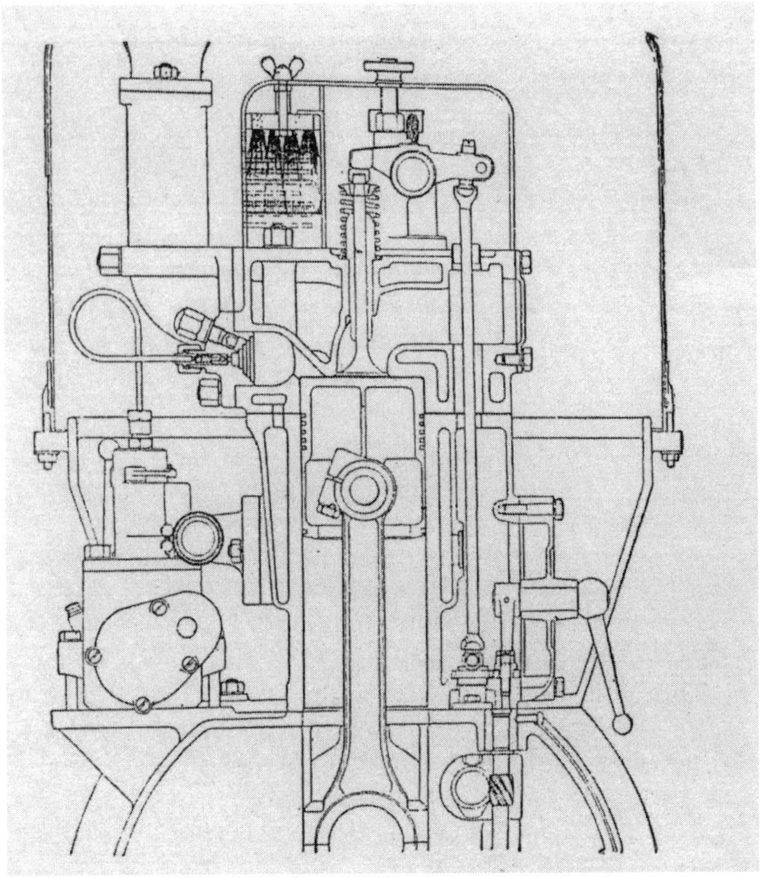

The diesel featured a pre-combustion chamber with decompression lever and overhead valves.

This survivor seems to be a transitional model between the petrol and full diesel types.

Austin diesel featured a geared-down handle on front of crankshaft or, as an expensive option, Bosch electric starter motor.

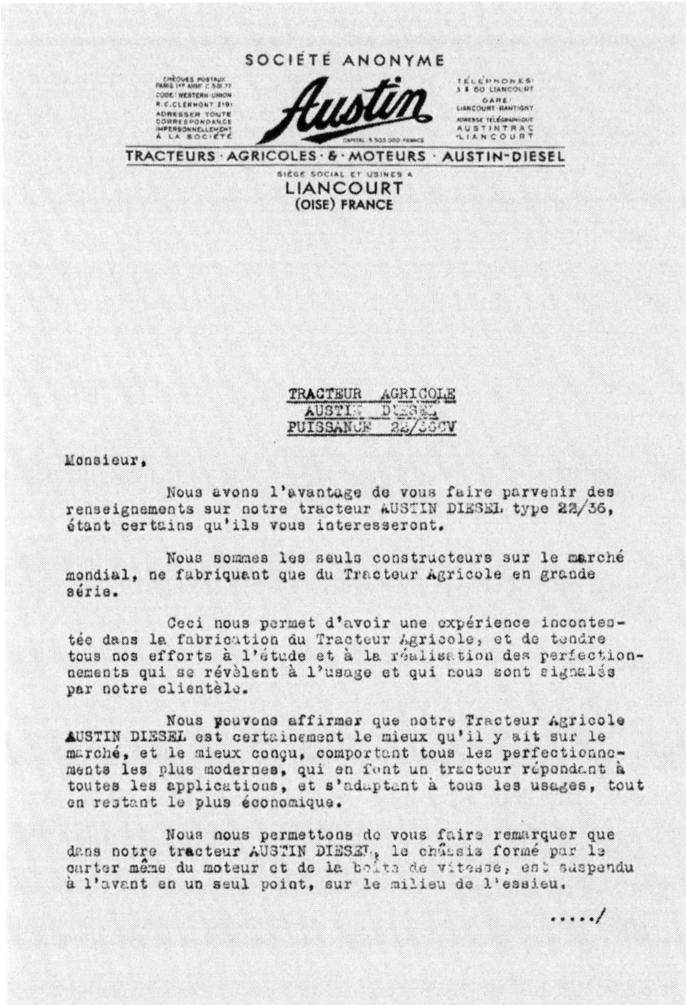

SOCIÉTÉ ANONYME

CHÈQUES POSTAUX PARIS 1er ARRT C 581.77
CODE: WESTERN-UNION
R.C.CLERMONT 2191
ADRESSER TOUTE CORRESPONDANCE IMPERSONNELLEMENT À LA SOCIÉTÉ

Austin

CAPITAL 5.500.000 FRANCS

TÉLÉPHONES: 3 & 60 LIANCOURT
GARE: LIANCOURT-RANTIGNY
ADRESSE TÉLÉGRAPHIQUE AUSTINTRAC-LIANCOURT

TRACTEURS · AGRICOLES · & · MOTEURS · AUSTIN-DIESEL

SIÈGE SOCIAL ET USINES A
LIANCOURT
(OISE) FRANCE

TRACTEUR AGRICOLE
AUSTIN DIESEL
PUISSANCE 22/36CV

Monsieur,

Nous avons l'avantage de vous faire parvenir des renseignements sur notre tracteur AUSTIN DIESEL type 22/36, étant certains qu'ils vous interesseront.

Nous sommes les seuls constructeurs sur le marché mondial, ne fabriquant que du Tracteur Agricole en grande série.

Ceci nous permet d'avoir une expérience incontestée dans la fabrication du Tracteur Agricole, et de tendre tous nos efforts à l'étude et à la réalisation des perfectionnements qui se révèlent à l'usage et qui nous sont signalés par notre clientèle.

Nous pouvons affirmer que notre Tracteur Agricole AUSTIN DIESEL est certainement le mieux qu'il y ait sur le marché, et le mieux conçu, comportant tous les perfectionnements les plus modernes, qui en font un tracteur répondant à toutes les applications, et s'adaptant à tous les usages, tout en restant le plus économique.

Nous nous permettons de vous faire remarquer que dans notre tracteur AUSTIN DIESEL, le châssis formé par le carter même du moteur et de la boîte de vitesse, est suspendu à l'avant en un seul point, sur le milieu de l'essieu.

...../

Part of a five-page sales letter signed illegibly by one of the directors. Note mention of motors in heading, which were available to re-power other tractors or for stationary roles.

This old Austin has been dieselised, though the motor is by Douge and not Austin.

Showing the geared-down hand starter that allowed adequate revolutions to be reached before disengaging the decompressors on the side. This is a 22/36 model preserved in the UK.

The bulkhead plate of a 22/36, numbered 9118. Whether Austin tractors were numbered in succession is unknown but they reached over 10,000.

The magneto confirms that this later pattern Austin is petrol-paraffin, yet the bi-bloc does not resemble the usual engine, which sits in an untypical subframe, so presents a mystery.

Another puzzling Austin, but this time too indistinct to make out more details than the hand starter.

A 22/36 hauling a Poclain side dump trailer.

One of the big diesels in Britain, where they were sold in small numbers 1934–40.

Another indistinct mid-1930s view but shown because in the background is the only part of the French Austin factory still standing in 2017.

A static Austin diesel engine for industrial or agricultural purposes. They came in 20/25, 30/35 and 50/56 ratings.

ECONOMICAL POWER.

Austin

45-55 H.P. DIESEL TRACTOR.

Modern farming requires the use of newer forms of power.

By greatly reducing the running costs, increasing the output, and the ability of being able to take advantage of all favourable weather conditions, the **45-55 H.P. AUSTIN DIESEL TRACTOR,** represents an enormous superiority over older and obsolete forms of farm power

The **45-55 H.P. AUSTIN DIESEL TRACTOR,** fitted with low pressure pneumatic tyres, can work on the land or travel on the highways without the extra work of changing over wheel equipment. Used in conjunction with a threshing outfit, or an agricultural trailer, it makes an ideal combination. For ploughing or other land work, the tyre treads dig in and give all the necessary grip on almost every type of soil.

Other wheel equipment consists of steel front and rear wheels, with quick detachable rear wheel rims fitted with efficient cleats, giving maximum adherence under worst conditions.

For further information, please address your inquiries to:—

In 1934 came this massive four-cylinder 7,200 cc 45/55, shown here in an English catalogue. It had six speeds and weighed 3.25 tons.

A preserved 45/55 in Britain, where the colour was no longer blue (as sometimes encountered in France) but green with red wheels. A chaff mesh protects the radiator core.

TRACTEUR AUSTIN-DIESEL, Type 22-35
CARACTÉRISTIQUES MOTEUR
4 Cylindres Monobloc.
Alésage .... 105 mm.
Course .... 27 mm.
Régime du Moteur .... 1200 TM
Vilebrequin à 5 portées.
Graissage sous pression.
Filtre à huile de graissage.
Refroidissement par Radiateur et Pompe.
Épurateur d'air.
Pompe d'injection Bosch.
Filtre à combustible Bosch.
Consommation par CV/H: 210 grammes.
AUSTIN-DIESEL

A French catalogue cover showing the 4.4 litre 22/35, as it was known in France.

This Austin 22/36 diesel is preserved in France. Paintwork is pale green, presumably based on original traces.

Buyers' guides don't list a 50/56 (threshing version to left) so this show picture is difficult to date, though is believed to be 1938.

Another show view with 22/35 and 45/55 visible and to the left of them Hanomags imported from Germany by R. Rothschild & Co.

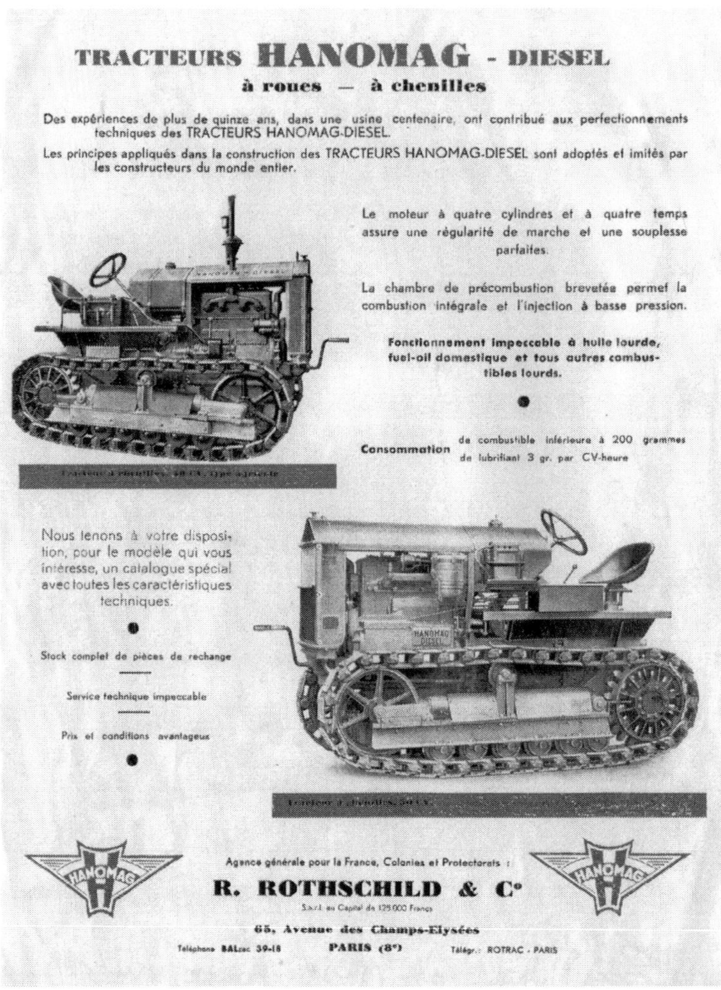

Robert Rothschild had experience of Hanomag before buying Herbert Austin's shares in SA Austin.

A 34 hp Austin (again a mysterious rating and conceivably the post-war prototype) with Austin staff including designer Gabriel Naninck second from left. Director Milos Celap is near the middle at rear with a Homburg hat.

Krupp stole the Austin factory and maybe used a Krupp lorry like this to transport its armaments produced by slave labour.

Longbridge returned to lorries in late 1938 and this is one of Austin's wartime K6 chassis equipped with Coles crane.

The return of peace to Liancourt with, naturally, an Austin pulling one of the floats.

This leaflet was handed out at a post-war agricultural show, though the style and lettering have a late 1930s appearance.

Herbert, First Baron Austin, died in May 1941, followed by his wife a year later. Their modest tomb is at Lickey Church.

Robert Rothschild and his helper Joanny Vincent were captured at Cléon D' Andran and executed in German-controlled concentration camps.

Plaque (which has since disappeared) on Austin factory remembering Robert Rothschild, Robert Pasquier, and Louis Fouquerolle, killed by the Nazis.

SFV and Austin French wrecks, a relatively complete early Austin at Dufresne's scrapyard at Villeperdue and an early UK Austin snapped at Redbourne, Herts, in 1949.

Many Austin Twelves ended up as rudimentary tractors during the Second World War.

In 2017 at least a dozen agricultural conversions of Austin Sevens exist.

A sophisticated rowcrop conversion of an Austin Seven by Wilkins brothers at Fladbury, Worcestershire, market garden demonstration in June 1945.

Homemade tractor in the Channel Islands features an Austin engine and gearbox with a Renault secondary gearbox.

Bristol crawlers were Austin-powered from 1942, well into the 1950s. This photo dates from 1947.

Showing the Bristol '20' Unit Construction.

Under the skin of an Austin-engined Bristol 20, the manufacturer having previously used Douglas, Jowett and Coventry Victor engines.

In the early post-war years some petrol-engined versions of Volvo tractors were British Austin-powered, though Volvo's purchase of the B-M diesel factory in 1950 ended this after a few years.

A 1949 Volvo advertisement celebrating the 10,091st tractor since 1946 – figures that SA Austin could only dream of.

The Ferguson from Austin's former Irish agent, Harry Ferguson, was the post-war tractor sensation. They were made in the USA, France (by Hotchkiss) and England (by Standard Motor Co.).

An hour north of Liancourt on the edge of Lille, the Canadian Massey-Harris Co. had been making farm machinery since 1926, adding small Pony tractors in 1951.

This could be a post-war Austin preserved in Britain. It is numbered 10010 and type 50D on its bulkhead plate while another to the side quotes type 100-10, order 709-51-31, constructeur presumé Austin. Maybe it was part of the 1950 tractor census referred to in the text when Austin was already a mystery.

In 1953 Massey-Harris bought Ferguson (here are some of their products at the time) and the combined Massey-Ferguson business built tractors at Beauvais from 1957.

1952 Newage advertisement for industrial version of Austin car engine used in Scottish-built Massey-Harris 726 combine. Note Bristol crawler and other applications.

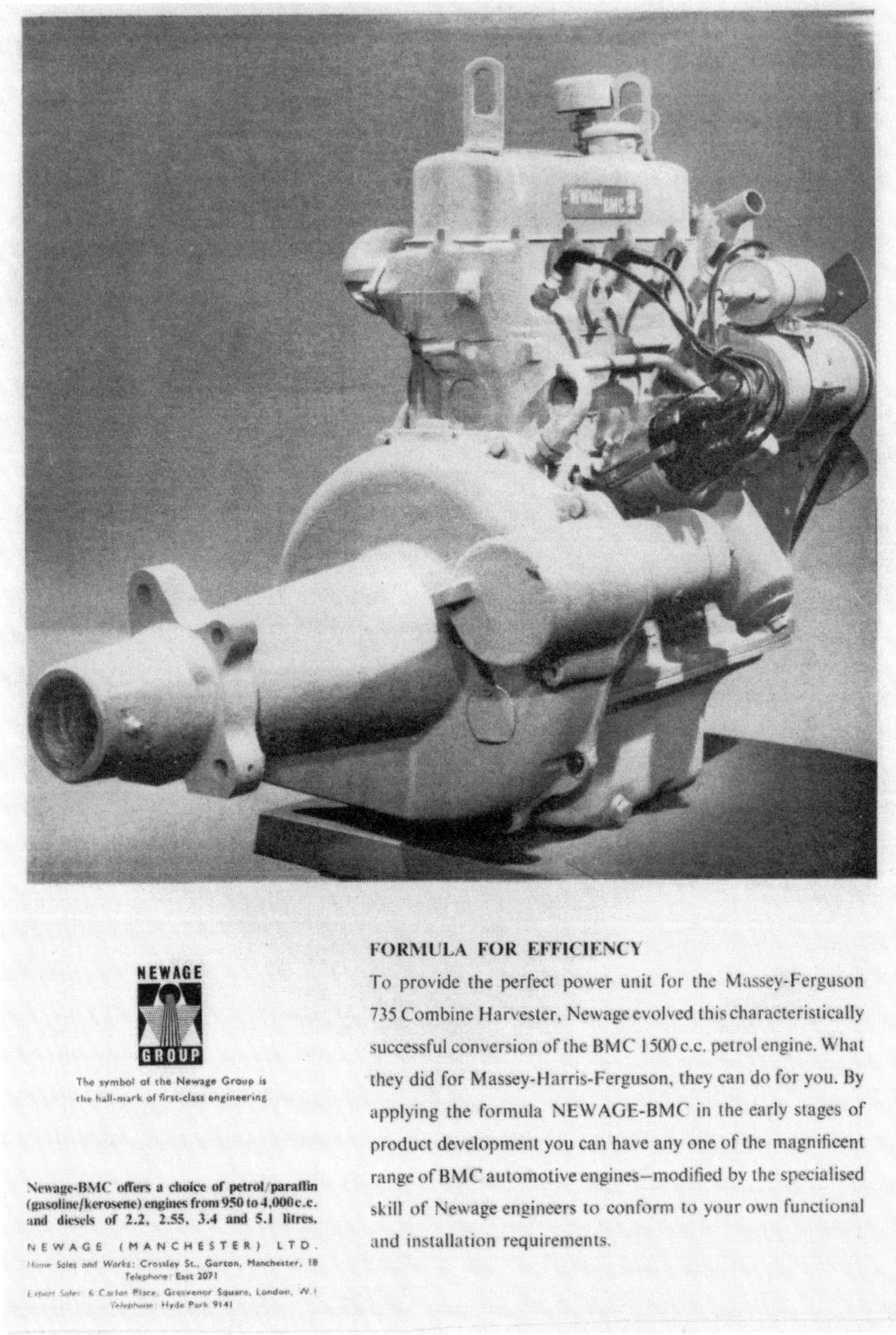

Following the 1952 merger of Austin and Morris (by then making Nuffield tractors) to create BMC, Newage continued in part of the old Crossley vehicle works to create special versions of their engines for M-F and others, and gearboxes for Winget and other small tractors.

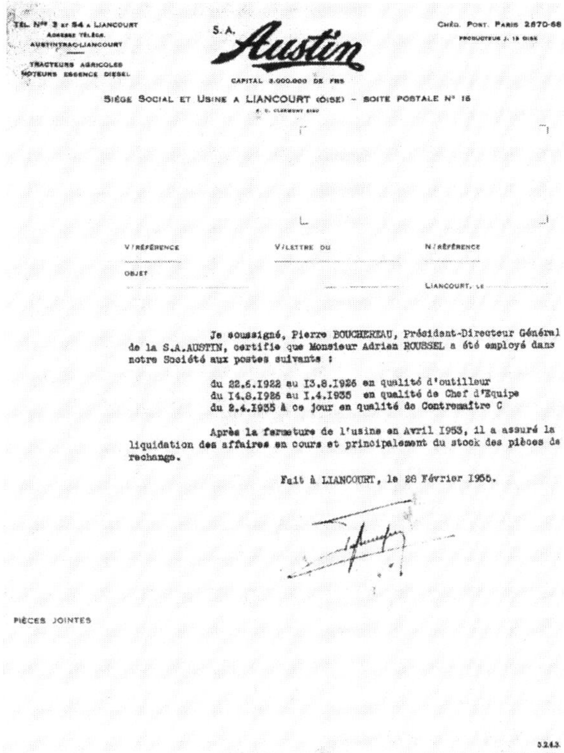

TÉL. Nos 3 et 54 à LIANCOURT
ADRESSE TÉLÉGR.
AUSTINTRAC-LIANCOURT

TRACTEURS AGRICOLES
MOTEURS ESSENCE DIESEL

S.A. Austin

CAPITAL 3.000.000 DE FRS

CHÈQ. POST. PARIS 2670-68

SIÈGE SOCIAL ET USINE A LIANCOURT (OISE) – BOITE POSTALE N° 15

V/RÉFÉRENCE

V/LETTRE DU

N/RÉFÉRENCE

OBJET

LIANCOURT, LE

Je soussigné, Pierre BOUCHEREAU, Président-Directeur Général de la S.A.AUSTIN, certifie que Monsieur Adrien ROUSSEL a été employé dans notre Société aux postes suivants :

du 22.6.1922 au 13.8.1926 en qualité d'outilleur
du 14.8.1926 au 1.4.1935 en qualité de Chef d'Equipe
du 2.4.1935 à ce jour en qualité de Contremaître C

Après la fermeture de l'usine en Avril 1953, il a assuré la liquidation des affaires en cours et principalement du stock des pièces de rechange.

Fait à LIANCOURT, le 28 Février 1955.

PIÈCES JOINTES

5243

Evidence that SA Austin still existed as a source of spare parts in 1955, though the works had closed in April 1953.

Austin's military Champ 4x4 was replaced by this civilian Gipsy in 1958 with 2.2 litre petrol or diesel engines.

Numerous types of crop sprayer are available, the one shown here being driven from a pulley on the centre power take-off. Other types that can be driven from the rear power take-off are also obtainable.
For the large farm or industrial undertaking—a mobile electric welding plant! Repairs can be dealt with immediately—on the spot—with this go-anywhere equipment installed in the vehicle and driven by the rear power take-off.
All in a day's work!
4X4
In its capacity as a general purpose vehicle the Austin Gipsy has many applications. Some examples of its versatility are illustrated here.
Another interesting job for the Gipsy is Using a drum type pulley driven by the rear power take-off, manual labour is reduced to a minimum with this stationary saw bench.
Portable saws are available for attachment to the Gipsy where mobility is the first consideration.
Many are the types of trailer that can be towed by the Gipsy. With four-wheel drive engaged, it has a maximum drawbar pull of 3,000 lb., sufficient to tow a heavy trailer over really rough, or heavy, ground.

Some agricultural applications for the Gipsy shown in a sales brochure, though the Land Rover remained more popular.

Aveling-Barford made graders under licence from the American Austin-Western from the 1940s and this is a 1970 Scania 275 bhp powered Aveling-Austin (no relation!). Aveling and Marshall were merged by British Leyland and adopted the Bristol crawler.

The Wolseley Sheep Shearing Machine business went through changes of name and ownership but still made farm equipment (1975 Wolseley-Webb mower shown) and later American yard tractors under the Wolseley name.

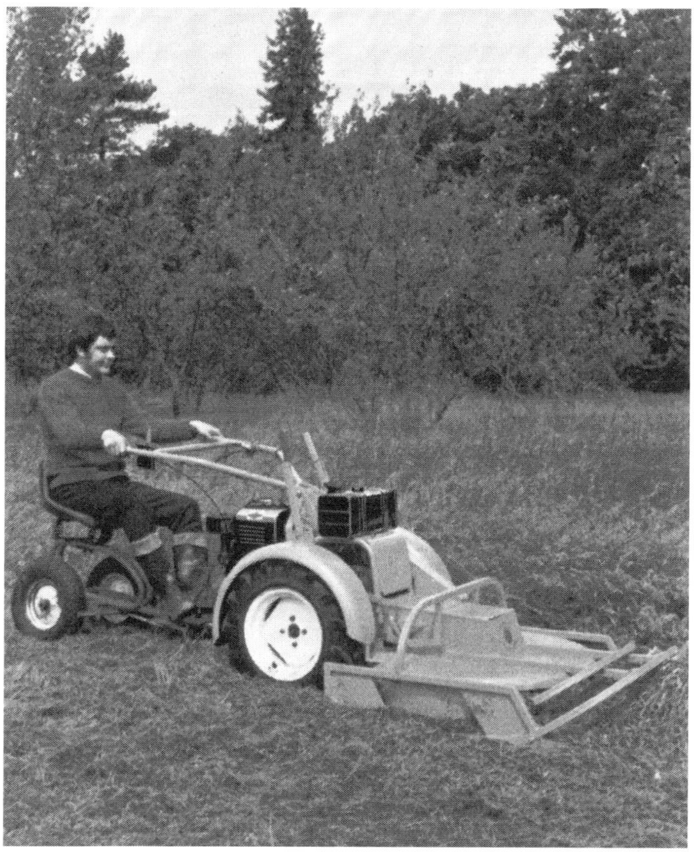

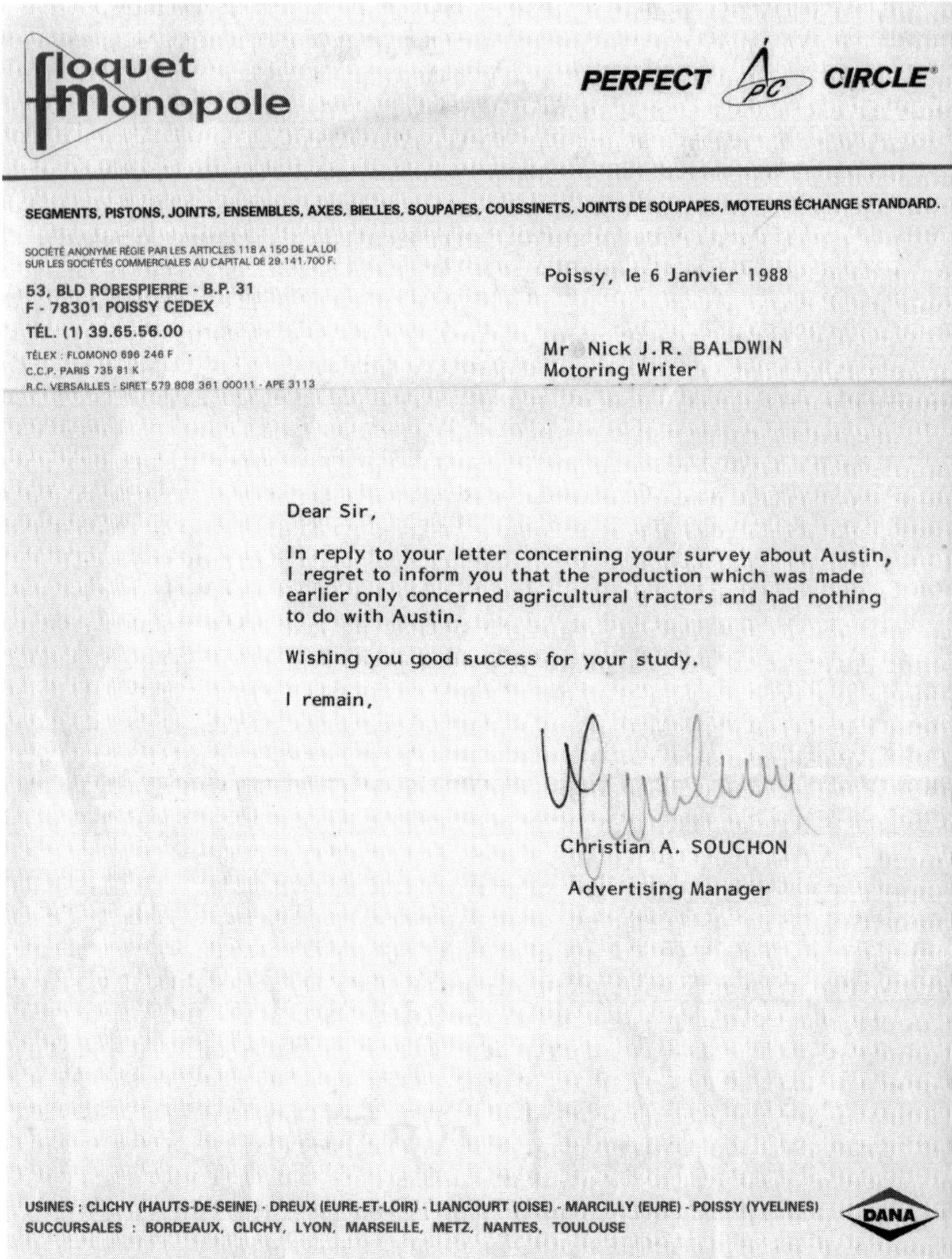

floquet Monopole

PERFECT CIRCLE®

SEGMENTS, PISTONS, JOINTS, ENSEMBLES, AXES, BIELLES, SOUPAPES, COUSSINETS, JOINTS DE SOUPAPES, MOTEURS ÉCHANGE STANDARD.

SOCIÉTÉ ANONYME RÉGIE PAR LES ARTICLES 118 A 150 DE LA LOI
SUR LES SOCIÉTÉS COMMERCIALES AU CAPITAL DE 29.141.700 F.

53, BLD ROBESPIERRE - B.P. 31
F - 78301 POISSY CEDEX
TÉL. (1) 39.65.56.00
TÉLEX : FLOMONO 696 246 F
C.C.P. PARIS 735 81 K
R.C. VERSAILLES - SIRET 579 808 361 00011 - APE 3113

Poissy, le 6 Janvier 1988

Mr Nick J.R. BALDWIN
Motoring Writer

Dear Sir,

In reply to your letter concerning your survey about Austin, I regret to inform you that the production which was made earlier only concerned agricultural tractors and had nothing to do with Austin.

Wishing you good success for your study.

I remain,

Christian A. SOUCHON

Advertising Manager

USINES : CLICHY (HAUTS-DE-SEINE) - DREUX (EURE-ET-LOIR) - LIANCOURT (OISE) - MARCILLY (EURE) - POISSY (YVELINES)
SUCCURSALES : BORDEAUX, CLICHY, LYON, MARSEILLE, METZ, NANTES, TOULOUSE

DANA

Floquet Monopole, which occupied Austin's Liancourt works in the 1950s–80s, plainly knew little about its illustrious history.

The only thing standing on the Austin site in 2017 is this tower that predates the tractors.

View across the cleared Austin site in 2003 to Bajac, which has since been demolished.

2017 and the Austin site remains unsold and is returning to nature.

Diesel Austin designer Gabriel Naninck's son, born 1939 (right), shows the author the slide rule his father used during his thirty years at the Austin works.

Printed and bound by CPI Group (UK) Ltd, Croydon, CR0 4YY

30/01/2025

01828412-0006